Saving Orchids

Saving Orchids

Stories of species survival in a changing world

Philip Seaton & Lawrence W. Zettler

Kew Publishing
Royal Botanic Gardens, Kew

The University of Chicago Press

© The Board of Trustees of the Royal Botanic Gardens, Kew 2025

Text © 2025 Philip Seaton and Lawrence W. Zettler
Images © as stated in the captions

The authors have asserted their right to be identified as the authors of this work in accordance with the Copyright, Designs and Patents Act 1988.

All rights reserved. No part of this publication may be reproduced, stored in a retrieval system, or transmitted, in any form, or by any means, electronic, mechanical, photocopying, recording or otherwise, without written permission of the publisher unless in accordance with the provisions of the Copyright Designs and Patents Act 1988.

Great care has been taken to maintain the accuracy of the information contained in this work. However, neither the publisher nor the authors can be held responsible for any consequences arising from use of the information contained herein. The views expressed in this work are those of the authors and do not necessarily reflect those of the publisher or of the Board of Trustees of the Royal Botanic Gardens, Kew.

First published in 2025 by Royal Botanic Gardens, Kew, Richmond, Surrey, TW9 3AB, UK www.kew.org
and The University of Chicago Press, Chicago 60637, USA

Royal Botanic Gardens, Kew

ISBN: 978 1 84246 831 9
eISBN: 978 1 84246 832 6

The University of Chicago Press

ISBN-13: 978-0-226-83941-7 (cloth)
ISBN-13: 978-0-226-83942-4 (e-book)

DOI: https://doi.org/10.7208/chicago/9780226839424.001.0001

British Library Cataloguing in Publication Data

A catalogue record for this book is available from the British Library.

Library of Congress Cataloging-in-Publication Data
Names: Seaton, Philip, author. | Zettler, Lawrence, author.
Title: Saving orchids : stories of species survival in a changing world / Philip Seaton and Lawrence W. Zettler.
Description: Chicago : The University of Chicago Press, 2025. | Includes bibliographical references and index.
Identifiers: LCCN 2024022126 | ISBN 9780226839417 (cloth) | ISBN 9780226839424 (ebook)
Subjects: LCSH: Orchids. | Orchids—Conservation. | Orchids—Climatic factors.
Classification: LCC SB409.48 .S43 2025 | DDC 635.9/344—dc23/eng/20240715
LC record available at https://lccn.loc.gov/2024022126

Design and page layout: Ocky Murray and Kevin Knight
Project management: Georgie Hills
Copy-editing: James Kingsland
Proofreading: Sharon Whitehead

Publication of this book was generously supported by the following donors:
Kit and La Raw Maran (Florida, USA)
The Chicago Botanic Garden (Illinois, USA)
The Venice Area Orchid Society (Florida, USA)

Page 2: *Odontoglossum crispum* growing in the Colombian cloud forest, with two seed capsules. Photo: Stig Dalström

Printed and bound in Italy by Printer Trento

For information or to purchase all Kew titles please visit shop.kew.org/kewbooksonline
or email publishing@kew.org

Kew's mission is to understand and protect plants and fungi, for the wellbeing of people and the future of all life on Earth.

Kew receives approximately one third of its funding from Government through the Department for Environment, Food and Rural Affairs (Defra). All other funding needed to support Kew's vital work comes from members, foundations, donors and commercial activities, including book sales.

Contents

Introduction 6
1 Ghosts in the 'land of flowers' 12
2 The crown jewels of the plant kingdom 26
3 The way we were 44
4 Orchids (almost) everywhere 60
5 Islands 64
6 Where are we most likely to find new species? 92
7 What's in a name? 96
8 Why we should care 110
9 Life (and death) in the anthropocene 122
10 Orchids on the edge 132
11 Who is doing what and where? 160
12 Where do we go from here? 178
13 Education: the teachers and the storytellers 186
14 Reserves 204
15 Reintroduction, restoration, and enhancement 220
16 Slippers 228
17 Living collections: safeguarding plants for the future 254
18 Seed storage 266
19 The way ahead 286
Further reading 294
Acknowledgements 294
Index 295

Introduction

'Humanity is waging war on nature. This is suicidal. Nature always strikes back – and it is already doing so with growing force and fury. Biodiversity is collapsing. One million species are at risk of extinction. Ecosystems are disappearing before our eyes. Human activities are at the root of our descent toward chaos. But that means human action can help to solve it.' António Guterres, UN Secretary General, 2021

(previous pages)
Cattleya trianae, Colombia's national flower, has been planted by schoolchildren in an avenue of trees leading into Guadalupe in Huila Province in the country's south.
Photo: Philip Seaton

This is a book of stories told by voices from around the world, stories about orchids and stories about some of the remarkable people who are working tirelessly to conserve these amazing and unique plants. An increasing number of people are becoming interested in orchid conservation. There are many groups globally, unsung heroes who are striving to conserve their own countries' floras. What gives us hope is the number of young people who understand that biodiversity matters. They are eager to play an active role in conservation.

Until recently our lives were enriched by a myriad of life forms that we assumed would still be here with us today. In parts of the US, listening to a night-time chorus of frogs in the neighbourhood marsh was a normal part of childhood. During the day, children would search for tadpoles, sometimes finding thousands wiggling among the aquatic plants along the shore, just steps away from lady's tresses orchids. Year by year, the chorus became quieter, and the marshes began to change. Today, only a few frogs and orchids remain. Is this the world we want our children to inherit? Do we want orchids and other life forms to slip through our fingers like tadpoles, eventually to vanish?

It's not that we can claim to have been unaware of the consequences of our actions. At the beginning of the 19th century, Alexander von Humboldt wrote about the disturbing changes taking place in the natural world in South America. In 1883, the director of Public Gardens and Plantations of Jamaica, Daniel Morris, wrote of his concerns about the impact of recklessly cutting down tropical forests. More recently, a steady stream of authors and colleagues have warned us about the biodiversity crisis now unfolding before our eyes. The human population has mushroomed over the past 70 to 80 years, and is predicted to peak at somewhere between 10 and 11 billion later this century. As a result, the loss of natural habitats has accelerated, and we have already lost a substantial portion of the world's biodiversity. We now live in an almost totally managed environment shaped by our own hands. We are the custodians of what remains.

Orchids are one of nature's 'canaries in the coal mine'. Particularly sensitive to subtle environmental changes, they provide us with an early warning system against potentially catastrophic environmental degradation. When these plants begin to disappear, even from seemingly pristine areas, it sparks a 'call to action', in much the same way that sudden widespread frog disappearances have garnered global attention. We begin with the story of Florida's iconic ghost orchid. Fewer than 1,000 plants remain, all clustered in remote swampy habitats in a region of the state at highest risk from sea level rise due to global warming. Until recently, the orchid has been difficult to cultivate, fuelling the appetite for its removal from the wild by poachers. Its allure is attributed to its lack of leaves, coupled with a stunning white floral display unlike that of any other orchid native to North America. Our story describes how college students and researchers, funded by a local orchid society, teamed up to demystify the ghost orchid, offering hope for its long-term survival.

Studying the past helps us to understand how things came to be the way they are today. It is like assembling a jigsaw puzzle in which many of the pieces are missing. By searching out and gathering the remaining precious fragments of correspondence from the early orchid hunters, accounts in journals such as *The Gardeners' Chronicle*, the occasional book and some grainy black-and-white images, a picture slowly begins to emerge of the natural world before it was radically changed by humankind. In the same way, Clyde Butcher's exquisite photographs of the changing landscape of Florida's Everglades over more than 35 years taken with a large-format camera reminiscent of those first developed in the 19th century, will provide an invaluable window into our world for future generations.

It is difficult to credit the numbers of orchids extracted from tropical forests by early collectors. Some species, such as *Odontoglossum crispum* – the 'Queen of Orchids' – were extracted in the hundreds of thousands, even millions. The Victorian era was one of 'orchid fever': wealthy European and American aficionados assembled enormous private orchid collections. What was the attraction then, and indeed what is the attraction today? Orchids have been described as 'the crown jewels of the plant kingdom'; the beauty of many species and the incredible diversity of their flower structure are unparalleled. Charles Darwin was besotted by orchids and their often bizarre pollination mechanisms. What is more, orchids are not confined to the tropics – Darwin studied orchids on a 'grassy bank' just around the corner from Down House, his home in southern England.

With currently around 30,000 named species and new ones being identified every year, the exact number of orchid species is still a matter of debate. Each new discovery needs a description and a name, and the science and art of taxonomy remain as important as ever. But why should we care? What does it matter if thousands of orchid species go extinct? The world would keep on turning. Aside from the moral argument – what right do we have to deprive future generations of the opportunity to enjoy the products of millions of years of evolution – there are practical reasons for conserving orchids. Some species contain biologically active compounds that may turn out to be important in treating cancers, for example.

We can't ignore the fact that many orchids are on the edge of an extinction precipice. Humans have been altering the natural environment since time immemorial. The story of the long-term deforestation and removal of the original vegetation cover of China over millennia provides an unusually well-documented example. Today it is difficult to conceive that there were once elephants as far north as what is now Beijing, and their long retreat from the north-east to the south-west mirrored modern China's economic development and environmental transformation. Five hundred years ago, Brazil's Serra dos Órgãos mountain range was completely

covered with dense, moist forest. Today only small fragments remain. The current and future impacts of global warming are particularly disturbing, but the story is far from being all gloom and doom. Researching this book has made us more optimistic about the future. Education about the natural world lies at the heart of conservation. Young people around the globe are increasingly urbanised, and have little opportunity to engage with the natural world. Yet motivated individuals such as Sergio Elórtegui Francioli in Chile remind us that, when students are taken into the field, they become hooked on the natural world for life.

Most tropical orchids grow high in the forest canopy, and Luciano Zandoná in the Atlantic Forest of Brazil and Rebecca Hsu in the mountains of Taiwan, share with us the exhilaration and excitement of climbing up into the canopy. Every living creature is connected to a host of others in a complex web of interdependence. When we attempt to save one orchid species from extinction, we must also include the living and non-living components of its ecosystem in one recipe for recovery. Many orchid species are protected in national and private reserves, and the good news is that more are being established. Reintroduction, restoration and enhancement all have their part to play. We highlight the plight of slipper orchids because, as a group, they include some of the most charismatic and (therefore) endangered species. In addition to conserving their habitats, maintaining orchids in living collections and as viable embryos in seed banks at universities and botanic gardens provides us with some insurance against future losses, safeguarding plants for the future.

Collected in vast numbers in the 19th century, remarkably *Odontoglossum crispum* (syn. *Oncidium alexandrae*) can still be found growing naturally in the cloud forests of Colombia.
Photo: Stig Dalström

1 Ghosts in the 'land of flowers'

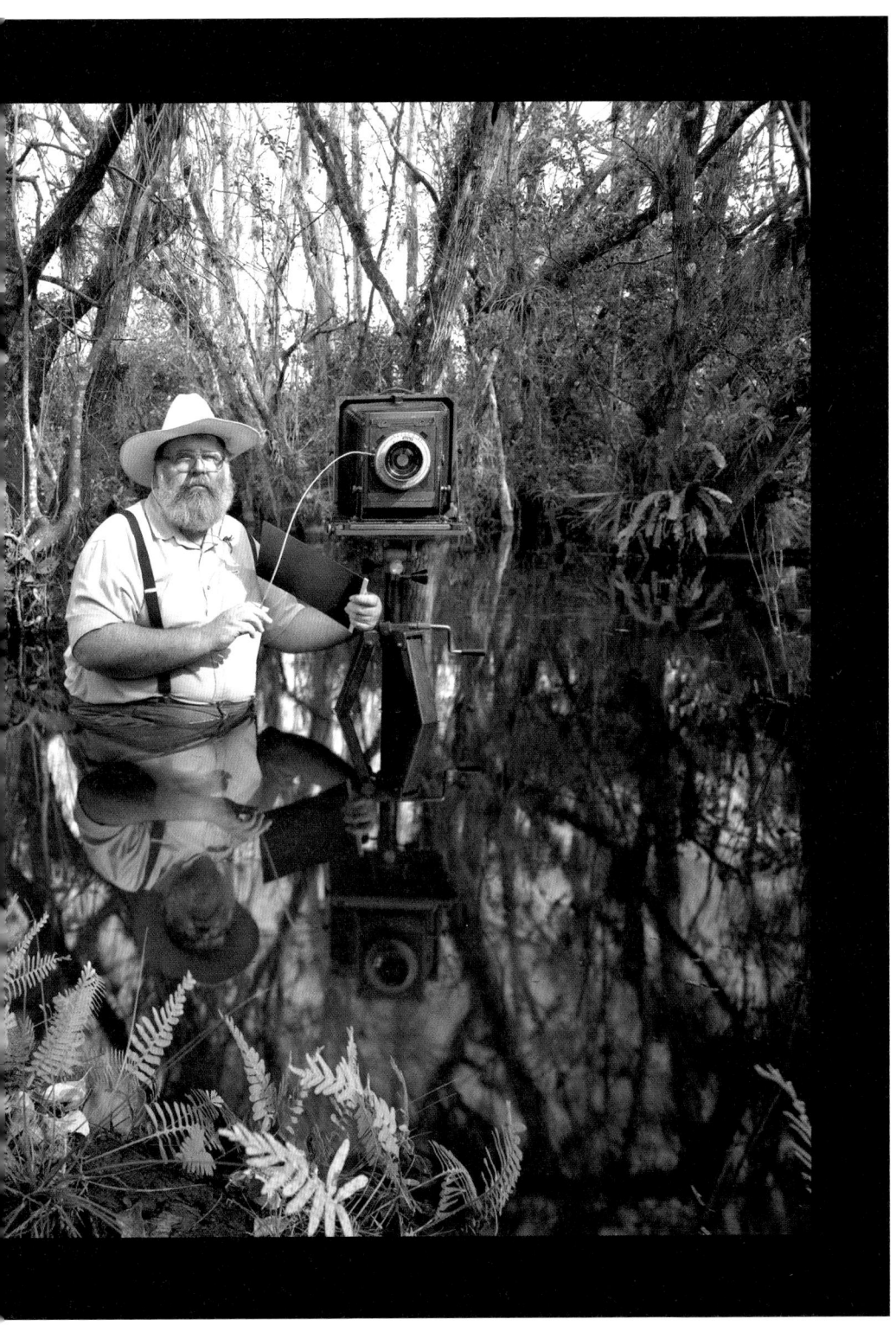

VIEWPOINT

Big Cypress Swamp
by Clyde Butcher

My first baptism in swamp water was in the Big Cypress Swamp in 1984, where I experienced a primeval presence I had never felt before. It was as though I was entering a time warp where the beginning and end of time combined in the waters of the swamp. It dawned on me, at that moment, that we humans are part of a greater whole. The ancient beauty of the cypress forest stunned me. With my eyes opened to the diverse beauty of Florida, I became aware of the need to get into it. However, the environment of the Everglades is intimidating. I worried about snakes, alligators, poisonous spiders, and just about anything else my imagination could conceive. How was I going to get the courage to hike out into waist-deep water to capture the images I wanted?

With the help of a good friend, Oscar Thompson, I began to adventure into the wilds of the Everglades. I learned that though the environment of the Everglades is unique, it is not something that should cause fear. As the population of Florida increased, and the beauty of the land was overrun by houses and strip-malls, I began to worry about the rapid disappearance of natural Florida. Suddenly, my photographs of the Everglades were becoming history rather than images of how Florida presently appeared. That knowledge was disturbing, and motivated me to begin my conservation activities. Saving the Everglades and working with state agencies and national environmental groups became part of my life, and the Everglades became the heart of my photography. Through my images, I hope

people will come to a greater understanding of the beauty we will lose if the preservation and restoration of our environment are not at the forefront of our thoughts. As stewards of the land, we must establish an ethic regarding the Earth that will lead the way toward a deeper and more personal connection to nature. As an artist, I reach deep within myself to express an image that will touch another. It is my sincerest hope that the vision I give to you of the Everglades will inspire you to love and protect our environment for generations to come.

Orchid diversity at the scene of an accident
'If you get orchids right, all other plants will follow.'
Kingsley Dixon, 2013

In south-west Florida lies the Fakahatchee Strand – an elongated, muck-filled pocket of shallow, tea-coloured water, shaded by a dome of trees crammed with reptiles. This unique and scary place, measuring a mere 2 by 35 km (1.2 by 22 mi), is North America's only orchid diversity 'hotspot'. Close to 50 orchid species have been documented here, almost a dozen of which are found nowhere else on the mainland, including Florida's crown jewel, the ghost orchid, *Dendrophylax lindenii*. Orchids and bromeliads crowd together on low-hanging tree branches, giving the Fakahatchee its nickname of the 'Amazon of North America'. This unique place has somehow evaded complete destruction in the modern age, serving as a portal into the past.

Florida epitomises the relationship between the harshness of the land, the people who have tried to tame it, and the orchid diversity it supports. In 1921, the Tamiami Trail blazed through the most treacherous stretch of south Florida, connecting east to west for the first time, and it was followed half a century later by a superhighway that still functions as the main land artery between Miami and Fort Myers-Naples. In the 1930s, large swathes of the Everglades – Florida's 'River of Grass' – were drained and diverted by the Army Corps of Engineers for the sake of agriculture, homes, and other forms of 'progress'. Ancient forests of cypress trees (*Taxodium distichum*) were felled in the mid-20th century, stripped of their orchids and bromeliads for the sake of rot-resistant lumber. These human activities took their toll. Today, south Florida is by no means the untouched wilderness that the Spanish explorer Juan Ponce de León experienced 500 years ago. But there are still just enough pockets and patches of forest left to remind us of what the past was like, and what the future may hold if the forest is restored. Studying these wild fragments, down to the smallest microscopic detail, will be pivotal in this century of radical change. South Florida is a place where humans and orchids have collided on the highway of progress.

The full impact of the collision began during the 1950s, when the state's population more than doubled to roughly five million.

(previous pages)
Photographer Clyde Butcher recording images of the Florida Everglades.
Photo: Woody Walters

(opposite)
A leafless orchid, *Dendrophylax lindenii* can be seen clinging to the bark of the host tree by its silvery roots. The plant pictured bears a long, thin, developing seed capsule.
Photo: Lawrence Zettler

One resident was Carlyle A. Luer, a physician who, along with his wife, developed a hobby for taking photographs of native orchids that began one day in 1957 when he encountered his first ghost orchid. Orchids are legendary for casting spells, and in Luer's case, this led to a 'gigantic effort to find and photograph every species of orchid known to have occurred naturally in Florida'. Year by year, he ventured into habitats throughout the state, photographing every orchid he could find. In 1972, his efforts culminated in the publication of *The Native Orchids of Florida*. Fifty years later, Luer's photos have immortalised how Florida once was. Luer was a perfectionist. Back in the days when cameras recorded pictures on film, he would spend as long as two hours photographing a single flowering plant as he attempted to capture the perfect image.

The case of the ghost orchid
The ghost orchid made its debut on an international stage when Susan Orlean's best selling 1998 novel, *The Orchid Thief*, first appeared on bookshelves, and then in 2002 as the Hollywood movie, *Adaptation*. Although the book was based on a true story about a poaching incident in the Fakahatchee Strand, the movie was largely fiction, portraying ghost orchid roots as being hallucinogenic when ground into a fine green powder and snorted like cocaine. For the record, no one can get high by snorting, smoking, or consuming 'swamp laudanum' from ghost orchid extracts, and there is no evidence that the plant has a place in alternative medicine. Let them be.

Of the estimated 1,000 wild ghost orchids thought to remain in Florida, it is not uncommon for a half dozen or more to fall victim in one year to the poacher armed with a hacksaw. Hidden trail cameras have been deployed in remote parts of the Fakahatchee to identify those who steal orchids – typically one or two men dressed in camo. In some instances, poachers will spot a trail camera and destroy it on the spot. The last images recorded by the camera will often reveal the offender with an angry facial expression so heated it could melt glass. One by one, these mature orchids, and the unique genetic code locked up in the nuclei of their cells, are whisked away out of the swamp, never to cross-pollinate another of their own kind again. Conservation is much more than keeping numbers up; it is ensuring that the rarest species can inherit the complete set of genes available in diverse individuals, which may be needed to survive in a rapidly changing world.

Political interference
The ghost orchid is not, however, confined to Florida, it also grows across the water in Cuba. In 2015, relations between the US and Cuba began to thaw, which soon led to scientists from both countries teaming up to conserve the same charming species. The first item on their agenda was to study the ghost orchid's habitat

Ghost orchid (*Dendrophylax lindenii*) in the Fakahatchee Strand.
Photo: Larry Richardson

Ernesto Mújica in the Florida Panther National Wildlife Refuge, July 2017.
Photo: Larry Richardson

to understand why the species was restricted in its distribution. After visas and funds were acquired, the researchers proceeded to collect data at the two sites that harboured the largest populations - Florida's Fakahatchee Strand and Cuba's Guanahacabibes National Park, located on the tip of the far western peninsula of the island.

The next step involved an ambitious effort to monitor all existing ghost orchids in the Florida Panther National Wildlife Refuge within the Fakahatchee Strand. Spearheaded by Cuba's Ernesto Mújica, the project's goal was to determine whether *Dendrophylax lindenii* populations in Florida were stable, increasing, or in decline. To build confidence in the mathematical models that would be generated, a minimum of five years of annual data was needed. Missing one year of data would result in the whole project being scrapped. Despite the risk, the decision was made to forge ahead.

Mújica's first trip to the US was in 2015. When he gazed into the Florida tree canopy, he was comforted by the sight of *Dendrophylax lindenii* and at least nine other orchid species that were also found in Guanahacabibes. In Cuba, he had become accustomed to losing blood to hordes of biting mosquitoes and to the sharp edges of karst limestone that scraped his feet and ankles, but not to the threat of being bitten by an alligator (*Alligator mississippiensis*) or a venomous cottonmouth snake (*Agkistrodon piscivorus*) that lurked in the shadows. In Florida, Mújica was accompanied by at least two college student interns, who acted as his personal assistants, and by at least one staff member whose primary duty was to spot large alligators. On one occasion, the group was forced to climb into a tree when an enormous alligator ventured a little too close for comfort. Despite dangerous encounters, heavy rain and frequent lightning, Mújica and the students were successful in acquiring a colossal amount of data during 2015, 2016 and 2017. The future looked promising.

After 2017, however, relations between the US and Cuba soured once again. They were made worse by the closure of the US embassy in Havana, which signalled an abrupt end to Mújica's direct involvement in south Florida, because he could no longer obtain a visa to enter the US. This sudden, unexpected development posed a direct threat to the project's completion. As the political situation worsened, it soon became clear that Mújica would not be able to step foot on US soil again any time soon.

Fortunately, during his last two trips to south Florida, Mújica had been assisted by Adam Herdman, a young college graduate from Illinois. He had taught Herdman how to spot tiny ghost orchid seedlings on tree bark, how to measure roots, and how to code and enter data. Herdman now returned to collect valuable Year 4 data on his own. With annual funding from the Naples Orchid Society, he ventured into the Panther Refuge each July for an additional three years, communicating with Mújica via e-mail. During July 2021, their efforts paid off when the much-awaited ghost orchid survey was published just in time to be included in a petition submitted to the US Fish and Wildlife Service to grant the orchid Federal protection.

The ghost buster

Prior to Mújica's arrival, only 16 individuals of *Dendrophylax lindenii* were known to occur in the Panther Refuge. During his first year, together with his student assistants, he recorded more than a hundred additional orchids, scattered in small, isolated pockets. Year by year, the numbers continued to rise. Even the smallest seedlings were no match for the keen eyes and experience of Mújica – the 'ghost buster'.

As of 2020, the Panther Refuge was home to about 400 ghost orchids, roughly 25% of the total number assumed to exist throughout Florida. The surveys revealed two highly significant findings. The first was that the more stagnant the water that is present beneath the host tree, the more orchids you will find in the canopy. Why? Possibly because the water creates a humidity dome capped by leaves and branches, insulating the orchids from winter cold snaps. The humidity dome may also create a moist environment conducive to the proliferation of mycorrhizal fungi needed for seed germination and seedling development. Of the seven populations surveyed, only one site harboured spontaneous seedlings. It was also the wettest. Was this a coincidence? Maybe not. In the 20th century, before wetlands were drained on a grand scale, ghost orchids were presumably much more common in and

A fig sphinx moth (*Pachylia ficus*), shown carrying pollinia (yellow) of the ghost orchid (*Dendrophylax lindenii*) on its head at the base of the proboscis. The photo was taken in the Florida Panther National Wildlife Refuge at 5:43 a.m. on 5 July 2018.
Photo: Carlton Ward

among low-lying cypress domes that flooded during the rainy season. It is conceivable that the largest specimens clinging to trees today germinated on branches decades ago, at a time when water levels were higher. What remains are a small number of relict water holes – most in the Fakahatchee Strand – where a few hundred mature ghost orchids cling to life in isolation. One by one, these individuals are succumbing to old age, severe hurricanes, and occasional poaching at a faster rate than they are being replaced. For all practical purposes, most ghost orchid populations are little more than retirement villages occupied by childless elders. Conservationists refer to orchids growing in habitats that harbour mature plants but lack the conditions needed to spawn spontaneous seedlings as 'senile populations'.

Considering that orchids have complex life cycles intimately linked to pollinators, mycorrhizal fungi, available moisture and other factors, determining the cause of the infertility is a tall task for those attempting to save an orchid species from extinction. Are ghost orchid populations senile? In Florida, the answer appears to be 'yes' in at least six of the seven sites that Mújica surveyed. These were the driest locations, so lack of moisture is the most likely cause of infertility, although other factors, such as a lack of pollinators, must be ruled out before jumping to conclusions.

The pollinator

During the pre-dawn hours of 5 July 2018, wildlife photographer Carlton Ward captured the first image of a ghost orchid pollinator in Florida. The image showed a small hawk moth species (*Pachylia ficus*) frozen in mid-air carrying pollinia (packets of pollen) affixed to the base of its feeding tube, or proboscis, as it hovered in front of the flower. Ward used a digital single-lens reflex camera that can detect rapid insect movement, even in darkness, using lasers that he aimed just above the flower's lip. Once the pollinator was detected, the camera automatically took a sequence of photographs in rapid succession lasting 1 to 2 seconds in total, while bright light flashes were beamed out into the darkness. Over a period of three years, other smaller hawk moth species were photographed pollinating or visiting ghost orchid flowers, filling a critical knowledge gap. Ward's camera helped to dispel a long-held assumption that the ghost orchid was pollinated by larger moths such as the giant sphinx moth, *Cocytius anteaus*. Thus, as far as ghost orchid sex is concerned, size does not seem to matter. To attribute the ghost orchid's reproductive problems solely to inefficient pollination and fruit set, therefore, may be a case of 'barking up the wrong tree'.

A fungal connoisseur

Lurking beneath the corrugated bark of the orchid's host tree is a world rarely seen by the human eye, a microscopic land populated

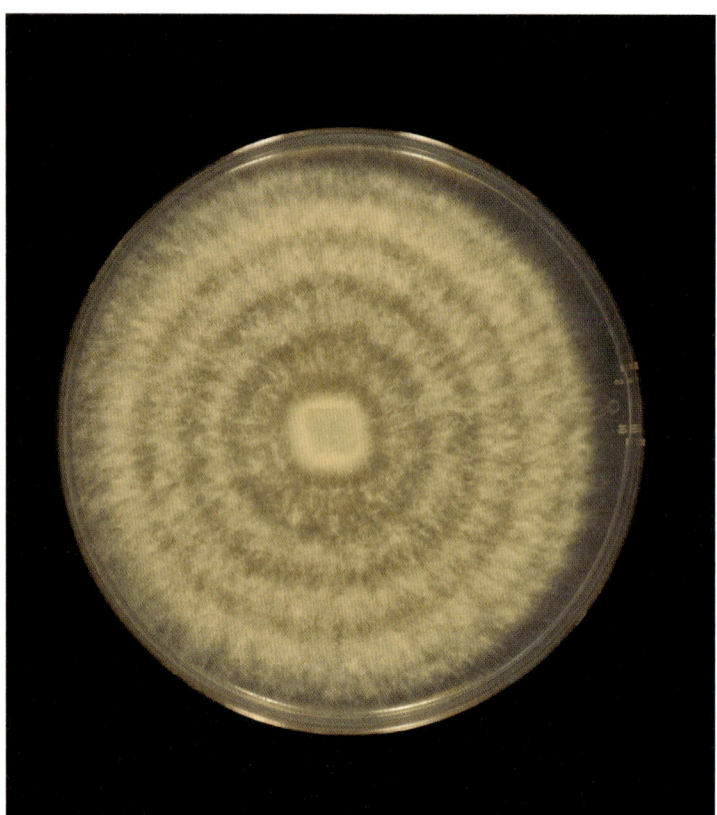

Ceratobasidium fungus growing on a special nutrient medium in a Petri dish.
Photo: Mike Kane

with bacteria, algae, lichen fragments, mites, and fungal threads (hyphae), all bathed in particles of organic matter and rainwater that trickles down from the tree canopy into the bark's cracks and crevices. In the Northern Hemisphere, the safest place for these organisms to live is in the folds of tree bark where direct sunlight is restricted for much of the day. For more than half of all ghost orchids surveyed in both Florida and Cuba, when their roots were traced back to a point of origin where the seeds had initially germinated, their birthplace was a crevice that faced north.

In nature, orchid seeds require the intervention of different kinds of fungi to germinate. Does the ghost orchid rely on a specific fungus? The answer appears to be 'yes'. When PhD student Lynnaun Johnson applied modern molecular techniques to ghost orchid populations within the Panther Refuge, he discovered that a species of *Ceratobasidium* fungus was widespread on tree bark and ghost orchid roots throughout the refuge. This was the case even in drier sites that lacked seedlings, not just at the wettest site where seedlings were known to occur. This suggests that the orchid's seeds were either not being effectively dispersed by wind, or that water levels were too low for seedling survival, and that ghost orchid populations therefore need higher water levels to overcome senility.

How can we be sure that *Dendrophylax lindenii* relies on this particular fungus? The answer was provided by students Nguyen Hoang and Ellen Radcliffe. While Johnson was side-stepping reptiles as he sampled trees and orchid roots for fungi, Radcliffe was draped in a white lab coat in the comfort of a small, air-conditioned laboratory. In July 2014, she isolated *Ceratobasidium* 394 from the ghost orchid. The fungus was shipped to Mike Kane's laboratory at the University of Florida, and into the care of Hoang for use in symbiotic germination experiments *in vitro*. After a few weeks' incubation with the seeds, the fungus outperformed all other germination treatments, including 'axenic' culture (in the absence of any symbionts), confirming that *Ceratobasidium* 394 was indeed *D. lindenii*'s mycorrhizal associate. When this fungus was analysed genetically, it turned out to be the same fungus that Johnson had detected in roots and on bark throughout the refuge. The ghost orchid is therefore a fungal connoisseur of sorts, targeting a specific fungus throughout its life.

Acclimatisation and reintroduction
One major hurdle remained that would decide whether the ghost orchid could be artificially propagated. Prior to Hoang's work, ghost orchids grown from seed in the laboratory had been available for sale. The origin of such seed was, and remains, sometimes questionable, but sales are strong, fuelled by the internet and those with deep pockets. However, many people have purchased seedlings only to have these delicate plants slowly decline and die under their watch, despite the best of care. To establish viable seedlings in a greenhouse therefore posed a formidable challenge, but solving this problem would potentially reduce the appetite for plants collected in the wild.

Hoang's experiments supplied this last critical piece of the ghost orchid's journey to recovery. Not only did his experiments confirm the significance of fungus *Ceratobasidium* 394, but he also cultivated thousands of seedlings in sterile culture. He carefully opened hundreds of plastic tissue culture vessels containing these seedlings, releasing them from captivity and exposing them to fresh air for the first time. In that instant, each seedling was immersed in a harsh world marked by drier air and peppered with the spores of microorganisms, including pathogenic fungi. It was Hoang's mission not only to keep the young orchids alive, but also to make them grow. This was the notorious Achilles' heel of the orchid's propagation, and he was determined to figure it out.

Hoang attached the seedlings to burlap – a brown, porous, biodegradable mesh made from vegetable fibres that can easily be affixed to host tree bark. He then either hung the burlap vertically, and misted the seedlings, or placed the fabric horizontally within a humidity dome. Vertical orientation and misting resulted in significantly higher survival. Not only did these orchids survive, most of them flourished. After three years, dozens flowered in a greenhouse at the University of Florida. They were suspended from the roof while rooted on burlap sheets fixed to wooden frames, and misted intermittently to keep the grey-green roots hydrated. Between 50 and 100 ghost orchids were attached to each sheet of burlap fabric, several of which flowered together at the same time. It was a breath-taking spectacle, a celebration that marked the end of the ghost orchid's long reluctance to grow anywhere outside of its swampy habitat.

The finishing line
Finally, numerous seedlings from the University of Florida were taken to the Panther Refuge for release. The fabric was cut into small squares, each containing one orchid, and reintroduced into suitable habitats for the first time. After attachment, the burlap eventually decomposes leaving no visible trace other than the seedlings that, by then, are firmly attached to the tree bark. Each square of burlap, with its precious orchid clinging by its roots, vaguely resembled a whole-wheat cracker ornamented by some exotic foodstuff. Piece by piece, orchid by orchid, several hundred were eventually affixed with the help of a staple gun to the bark of either pop ash (*Fraxinus caroliniana*) or pond apple (*Annona glabra*). Here, they would be monitored yearly, like their wild parents, in a place where they were meant to thrive.

Another 100 ghost orchids were also taken to the Naples Botanical Garden, where they were placed on public display along a board walk that wound in and around a young cypress dome. In this manner, science and education were locked in a mutual symbiosis, setting the stage for a new generation of caretakers to learn about, and appreciate, one of the loveliest orchids of the Northern Hemisphere.

Ghost orchids (*Dendrophylax lindenii*) established ex vitro on burlap suspended vertically, three years after they were germinated from seed in vitro. Photo: Mike Kane

2 The crown jewels of the plant kingdom

'The forest is a vast laboratory in which new species are produced, tested, and eliminated if found defective.' Alexander Skutch, 1971

Every living creature that flies, swims, wiggles, or takes root is the result of evolution – a genetic transformation in a population of organisms resulting in modifications that favour their survival. These transformations are the product of advantageous chance mutations that accumulated over millions of years. Orchids reign supreme as evolutionary warriors that make up around 10% of all the world's flowering plants. Why are they so successful? What makes them so different from other plants? Answering these fundamental questions holds some of the keys to their conservation and survival.

The flower

'I touched the antennae of Catasetum callosum *whilst holding the flower at about a yard's distance from a window, and the pollinium hit the pane of glass, and stuck by its adhesive disc, to the smooth vertical surface.'* Charles Darwin, 1862

Darwin was besotted by orchids and their many bizarre pollination strategies. Three years after publishing *On the Origin of Species*, he published *On the Various Contrivances by which British and Foreign Orchids are Fertilized by Insects*. Today, orchid pollination remains a fascinating and important topic of study, with new discoveries being made each year, propelled by the advent of new technology. In Yaoundé, Cameroon, the botanist Vincent Droissart and his team have started to develop and test a new camera that can film orchid pollinators in the canopies of tropical rainforests, and they have published a practical guide on how to build and use it. Surely Darwin would have loved today's digital camera traps.

Sexual reproduction unleashes genetic variation that serves as the foundation for rapid evolution shaped by natural selection. But when it comes to sex, animals hold a major advantage in that they can move freely from one location to another, actively seeking mates, whereas plants are stuck where they are rooted. Early in their evolution, plants had to rely on wind and water for pollination, so the appearance of the first flower was a game-changer that coincided with the diversification of winged insects. When flowers opened their petals for the first time, advertising and flaunting their pollen and sugary nectar as food, the days of haphazard sex were over. Flowers levelled the playing field by manipulating tiny animals to serve as mobile sexual organs. The first insect pollinators were probably beetles. Messy eaters, they didn't distinguish between pollen and nectar. Beetles were also cumbersome flyers that were lucky even to get off the ground. Orchids soon began to evolve flower parts that allowed them to divorce these pollinators, with their bad table manners and

(previous pages, clockwise from top left)
Ophrys tenthredinifera (Mallorca), *Cypripedium flavum* (China), *Angraecum sesquipedale* (Madagascar), *Cuitlauzina pendula* (Mexico), *Zygopetalum maxillare* (Brazil), *Prosthechea cochleata* (Guatemala), *Lepanthes dougdarlingii* (Colombia) and *Chloraea disoides* (Chile).
Photos: Philip Seaton, Holger Perner, Philip Seaton, Philip Seaton, Luciano Zandoná, Mayra Maldonado, Sebastian Vieira and Sergio Elórtegui Francioli

(opposite)
Epiphytic orchids (in this instance *Laelia anceps*) can often grow into large colonies.
Photo: Andrés Ramos

reluctance to fly, and instead team up with bees, butterflies and moths. They pulled this off by transforming their flowers into an irresistible visual and olfactory advertisement. One of the petals was modified to become the lip (labellum), which acted as a colourful flag to attract the attention of the pollinator, and as a landing platform for visitors. Over time, orchids began to target specific groups of winged pollinators by tweaking their floral displays, colour by colour, shape by shape, and chemical by chemical, evolving a bewildering array of often bizarre mechanisms to ensure pollination. Unfortunately, there can be a downside for orchids that target specific pollinators: they are more vulnerable to extinction should their pollinators disappear.

Orchid pollen is clustered into lightweight packets (pollinia) that are concealed behind the anther cap on a unique projection in the centre of the flower called the column. Pollen is 'glued' to the insect visitor, usually somewhere on the head (often the face or compound eyes) or on top of its back between the wings. The aim is to entice the pollinator to visit another flower of the same species and transfer the pollen to its stigma, a sticky liquid patch located beneath the column behind the anther cap.

Some flowers emit delicious perfumes: *Cattleya maxima* has a typical heady 'orchid perfume'; *Maxillaria tenuifolia* smells of coconut; and *el pelícano*, the green swan orchid (*Cycnoches chlorochilon*), smells of banana popsicles. Bulbophyllums often have dull, reddish flowers that release a subtle, but foul, stench like that of rotting meat that attracts flies. For night-pollinated species, powerful fragrance takes precedence as the primary cue for attracting pollinators on the wing, such as moths. Combining gas chromatography with mass spectrometry allows researchers to determine the blend of chemicals present in orchid fragrances. When the fragrance of *Dendrophylax lindenii* was analysed, eight compounds were discovered that were thought to attract moths. Mixed in the lab in the correct proportions, the blend of chemicals mimicked the sweet smell of the flower.

Fragrance is one ploy that orchids use to attract the attention of pollinators. The second is visual. Orchid flowers that attract night-flying moths are typically white, whereas brightly coloured flowers generally cater to bees, butterflies, and other day-flying creatures. The combination of sight and smell are often irresistible, and pollinators soon find themselves at the entrance of the open flower itself. For many orchids, nectar serves as the ultimate reward, providing a

> In 2007, the discovery in the Dominican Republic of a bee entombed in amber, dated to between 15 and 20 million years ago, with a package of orchid pollen on its back created a sensation in the orchid world. Ten years later, a fungus gnat carrying orchid pollen, found in a piece of Baltic amber from 45 to 54 million years ago, pushed the origin of the orchid family even further back in time. Advanced DNA-sequencing techniques now suggest that orchids first appeared over 100 million years ago during the mid-Cretaceous, when dinosaurs ruled the Earth, and winged insects were beginning to diversify.

Students collecting the floral fragrance of the ghost orchid, *Dendrophylax lindenii*.
Photo: Larry Richardson

sugar-rich meal needed for powered flight. The key to the orchid's success is placing the nectar deep within the flower. To gain access, the pollinator must work for it by inserting its mouthparts into a pouch or spur. The deeper it probes, the more likely it is that the pollinator will make physical contact with the orchid's sticky pollen mass. After securing a sugar meal, it exits the flower refreshed and nourished, and flies off to visit the next one.

Numerous studies have confirmed the importance of nectar for effective pollination and reproductive success in orchids, but its chemical composition has received surprisingly little attention. Recently, however, the ghost orchid's nectar was analysed in Florida for the first time and revealed sugars (glucose, fructose, sucrose), three acids (lactic, malic, threonic), as well as 4-hydroxyl benzyl alcohol. Some researchers suspect that pollinator mouthparts may introduce microbes into nectar tubes that could alter its chemical composition through fermentation.

Many of Madagascar's Angraecoid orchids have extremely long nectar spurs. Darwin surmised that they must be pollinated by moths with equally long tubular mouthparts. Aptly named 'Darwin's orchid', the most notable example was *Angraecum*

sesquipedale. A close relative of *Dendrophylax lindenii*, the nectar spur of *A. sesquipedale* reaches an incredible 40 cm (15 in) in length. When this orchid was brought to Darwin's attention, he hypothesised that someday someone would capture the pollinator, which would have a proboscis the same length as the nectar spur, a prediction that was confirmed 21 years after his death – the pollinator was a hawk moth with an enormously long tongue.

The flowers of *Disa uniflora*, the Pride of Table Mountain, are pollinated exclusively by the mountain pride butterfly (*Aeropetes tulbaghia*), which is attracted by their large, brightly coloured carmine sepals. Some orchids, like the diminutive members of the sub-tribe Pleurothallidinae, target tiny fungus gnats. On Réunion Island, the raspy cricket (*Glomeremus* species) pollinates *Angraecum cadetii*. In Japan, thrips are suspected of pollinating the terrestrial orchid *Epipactis thunbergii*. The flowers of *Cypripedium subtropicum*, from south-west China, mimic aphid colonies, they produce chemicals that attract hoverfly pollinators seeking the honeydew excreted by the insects. A large proportion of orchids found on the windswept Arctic tundra are pollinated by mosquitoes. In the tropical Americas, *Elleanthus* are typically pollinated by hummingbirds. In the Indian Himalayas, the red flowers of *Cryptochilus sanguineus* are pollinated by sun birds. There is even one report from China of a mammal, the wild mountain mouse, *Rattus fulvescens*, transferring the pollinia of *Cymbidium serratum* while feeding on the labellum of the orchid.

Some species have resorted to a life of celibacy by having self-pollinating flowers. Others, such as Australia's hammer orchids (*Drakaea* species), went in the opposite direction by customising

(above left)
Moisture present on the labellum of a ghost orchid (*Dendrophylax lindenii*) is collected using a glass micropipette for chemical analysis.
Photo: Adam R. Herdman

(above right)
Nectar is collected from within the spur of a ghost orchid (*Dendrophylax lindenii*) using a glass micropipette.
Photo: Adam R. Herdman

(opposite top)
In Chile, male *Colletes cyanescens* bee with pollinia of *Bipinnula fimbriata* 'glued' to its thorax.
Photo: Sergio Elórtegui Francioli

(opposite bottom left)
Disa uniflora in South Africa is pollinated by the mountain pride butterfly.
Photo: Philip Seaton

(opposite bottom right)
Elleanthus lupinus in Colombia is pollinated by hummingbirds.
Photo: Philip Seaton

their flowers to mimic gravid female wasps, catering to sexually aroused male wasps that attempt to mate with the flower in a process known as 'pseudocopulation'. Instead of passing on wasp genes to the female of its own kind, male wasps receive orchid genes in the form of pollinia and are apparently oblivious to the orchid's trickery as they fly off to another flower, driven by promiscuity. In south-east Australia, different populations ('ecotypes') of the hammer orchid, *Drakaea livida*, attract different male thynnid wasp pollinators by producing different pheromones (chemical signals). Such populations may gradually diverge over time and become separate species.

Be fruitful and multiply
'We cannot discount the likelihood that one in 10 billion seeds could be blown from Africa to South America in a viable condition.'
Robert Dressler, 1981

The first orchids were probably terrestrial and lily-like, bearing colourful fleshy fruits that may have been eaten by the ancestors of today's birds. A glimpse into the past can be seen in Asian forests where *Neuwiedia nipponica* produces several red, fleshy fruits when ripe. These fruits attract birds which disperse the seeds after they have passed through their guts. *Apostasia nipponica* has inconspicuous green fruits that are dispersed by crickets. Seeds of *Vanilla* are hard, black, and lustrous, and dispersed by as yet unidentified animals - perhaps small rodents.

Thousands of pollen grains may fertilise thousands of ovules that, in turn, result in thousands of seeds, each containing an embryo sired by a different sperm from a different pollen grain, and therefore each containing a unique genetic blueprint. Soon after pollination, the ovary begins to swell, forming an oval or elongated fruit or 'pod' (botanically, a capsule). The largest are about the size and shape of a lemon, and the smallest could fit within a peppercorn. Orchid capsules are dry when ripe and split open releasing many thousands, and in some cases more than a million, dust-like seeds into the breeze. Darwin found that a single capsule of *Orchis maculata* contained 6,200 seeds, or 186,300 on a single plant. Allowing for 400 non-viable seeds in a capsule, he estimated that, 'the possible rate of increase of this species is such that the great grandchildren of a single plant would nearly clothe with one uniform green carpet the entire surface of the land throughout the globe.'

Orchid seeds are not just tiny, they are essentially microscopic, often measuring fractions of a millimetre in length. Each seed typically contains an embryo comprised of around 120 cells. Enclosed within a papery thin seed coat (testa) and buffered by a small air space, they are 'balloon seeds'. The dead, empty cells of the testa form a sculptured lattice resembling a honeycomb that can catch the merest breath of air. Darwin noted that: 'The minute seeds within their light coats are well fitted for wide dissemination: and I have several times observed seedlings springing up in my orchard and in a newly-planted wood, which must have come from a considerable distance.'

Most orchid seeds, nevertheless, probably fall a short distance from the parent plant. Seed dispersal of some European orchids, for example, usually occurs within a few tens of centimetres (8 in) of the parent. Under the right circumstances, however, orchid seeds are capable of travelling far greater distances. Immediately after the volcanic island of Krakatoa erupted in 1883, it became completely sterile. Even so, just 13 years later, investigators found *Arundina*, *Cymbidium* and *Spathoglottis* growing on the island. By 1933, they had found 18 species of terrestrial orchids and 17 epiphytic species. Surveys conducted between 1991 and 1998 have since found more than 63 orchid species that must have travelled 40 km (25 mi) from nearby islands, including Java and Sumatra. Similarly, the fungal associates needed for seed germination must also have found their way onto the island, presumably via wind-borne spores. In the New World, cyclones (hurricanes) may have seeded south Florida with orchid seeds from tropical epiphytes rooted on trees in the Caribbean.

Counting seeds

We are told that a single seed capsule of *Grammatophyllum speciosum* (the Asian 'tiger orchid') may contain more than two million seeds. Who has counted them? How have they been counted? One at a time? Clearly not. When Suphat Rittirat and Sainiya Samala, two PhD students from the Prince of Songkla University in Thailand, visited the Millennium Seed Bank at Wakehurst (Kew's wild botanic garden in Sussex) to learn more about orchid seed storage, they brought with them two ripe capsules of *G. speciosum*, each resembling a large lime. One of the capsules was opened, and its contents weighed. Orchid seeds are so light that even two million seeds weigh very little, so that a microbalance which is accurate to five decimal places is needed. A small sub-sample of seed was weighed and the number of seeds in the sub-sample counted using a binocular dissecting microscope. It was then a simple matter to calculate the total number of seeds in the capsule.

Not all species, however, contain enormous numbers of seeds per capsule. The tiny fruits of pleurothallids and *Lepanthes* contain hundreds rather than thousands of seeds, and for collection purposes it is therefore necessary to harvest seeds from a number of capsules, and from separate individuals where possible. These differences in seed quantity between species reflect each orchid's ecological strategy for maximising its survival in the highly competitive natural world. Thus, each orchid tells its own unique life story, and seed number is merely the first chapter.

Weedy orchids

Some orchids are distinctly 'weedy' in that they colonise freshly disturbed sites and soon sprout in high densities. They also appear in unexpected places. A classic example was the appearance of large numbers of early spider orchids (*Ophrys sphegodes*) on the chalk spoil at Samphire Hoe in Kent, which had been excavated during the construction of the Channel Tunnel. Weedy orchids can be native species, but many have escaped captivity or entered a country undetected as stowaways on foreign cargo. In Florida, the lawn orchid (*Zeuxine strateumatica*) fits the latter category as a hitchhiker in ornamental centipede grass (*Eremochloa ophiuroides*), probably shipped from southern China. In Cuba, *Phaius tankervilleae*, a species native to many Asian countries, colonises clay tile roofs in towns, adding a touch of colour when they flower *en masse*. Also native to Asia, *Eulophia graminea* can be found sprouting up between cracks in pavements (sidewalks) in Florida.

The thought of orchids colonising our landscape may seem appealing, but it can be a serious ecological problem, especially on islands. In Hawaii, 14 species or hybrids have become naturalized, including *Dendrobium antennatum* from New Guinea and tropical North Queensland, and *Polystachya concreta*, a species found in America, Africa and Asia. As the botanist Jim Ackerman noted, these orchids have succeeded in 'breaking the chains of

The tiger orchid, *Grammatophyllum speciosum*, is distributed throughout many countries in South-East Asia, including Singapore and Thailand.
Photo: Philip Seaton

horticultural servitude and have made a successful run at freedom'. At a recent conservation conference, Ackerman solicited feedback from the audience as to why more was not being done about invasive orchids. After a long silence, one member commented: 'Uhhh, well, Jim, it's because they are orchids.'

Mycorrhizal fungi

Why are some orchids weedy whereas others are so rare? The answer, in part, lies in a cryptic world of death and decay ruled by fungi. These misunderstood but important organisms are ubiquitous and highly diverse – they live in the soil, on tree bark, and even in the ocean. They decompose dead organisms into smaller molecules that plants can then reabsorb. Fungi are not immune to being eaten, however, as they are filled with carbohydrates of their own, as well as water and mineral nutrients. Millions of years ago, orchids figured out how to feast directly on fungi through their roots, supplementing the food gained from their leaves through photosynthesis. Yes, that is correct, orchids are plants that feed upon fungi, not vice versa, and this adaptation is believed to have added to their evolutionary success.

Their appetite for fungi begins at birth in their dust-like seeds. Having sacrificed the luxury of a food reserve, orchid seed embryos utilise fungi to initiate germination. But not any fungus will do. Some orchid embryos prefer a restricted diet, feeding on one specific type of fungus needed to trigger germination, as in the case for Florida's ghost orchid (see page 22). Embryos of other orchids, especially 'weedy' species, are far less discriminating, and will consume several different kinds of fungi. This explains why weedy orchids sprout in unusual places, sometimes continents apart.

The challenge faced by all orchids is to release their seeds into the air and hope that at least one makes physical contact with a suitable fungus upon landing. The more seeds are produced per capsule, the more likely it is that one will reach its target. The odds are slightly higher for 'weedy' orchids because they have a more varied fungal diet. Once the embryo begins to feed on the fungus, it enlarges, ruptures its seed coat, and forms a spherical leafless body termed a protocorm. Eventually, the protocorm initiates leaves and begins to feed itself by photosynthesis. At maturity, orchids with well-developed leaves rarely grow tired of feasting on the fungi trapped within their roots, which allows them to supplement the sugars they make from photosynthesis with food taken from the fungi in a 'parasite-provider' relationship. This dual system of feeding distinguishes orchids from all other major plant groups that enlist the help of fungi as mutual partners. Simply stated, orchids apparently harm their fungi by consuming them, giving little if anything back in return. In doing so, however, orchids have eaten themselves into a corner where they now depend entirely on fungi for their survival.

Seed packets

The concept of sowing orchid seeds outdoors seems absurd because they are simply too small. How could anyone find them after they had been sprinkled over a substrate, or monitor their germination over a period of years? In 1993, Hanne Rasmussen and Dennis Whigham crafted an ingenious technique for capturing orchid fungi in soil using photographic slide mounts, thereby revolutionizing our understanding of seed germination in nature. Orchid seeds are sprinkled into a square of nylon netting, which then inserted into a 35 mm slide mount. Finally the mount is stapled shut and buried in soil. After one or two years, fungi growing in the soil slip through the microscopic pores in the nylon mesh and into the seeds. Because nylon is resistant to decay and the pores are too small to allow the seeds to escape, early stages of germination can be documented once the 'packets' have been retrieved, opened, and observed using a dissection microscope. Fungi implicated in natural seed germination processes are thereby captured and can be isolated from any 'protocorms' – tuber-like masses of cells produced by orchid seedlings in association with mycorrhizal fungi just before they produce the first leaf. Numerous researchers have since used this technique successfully for terrestrial orchids, and it is now being applied to epiphytic orchids in tropical regions by researchers willing to climb into trees, who 'sow' the packets on bark.

Life as an epiphyte

'Since orchids never die, unless by accident, and never cease to grow, there is no limit to the bulk they may attain. Mishap alone cuts their lives short – commonly the fall or the burning of the tree to which they cling.' Frederick Boyle, 1900

Part of the success of orchids lies in their epiphytic lifestyle, growing on the trunks and branches of trees. Despite a widespread misconception that orchids are parasites of trees, this is not the case. Trees simply provide mechanical support. The downside is that the lifespan of any given epiphyte is limited by the lifespan of the host tree it grows on. A traveller walking through a tropical forest will find an abundance of orchids clinging to broken branches that have fallen from above due to storms, wind or the tree's old age. These plants are doomed to a slow death by starvation, as they can no longer capture enough sunlight needed to generate food (sugar) from photosynthesis. Some fallen orchids may linger longer than others on the dim forest floor, especially those with roots harbouring mycorrhizal fungi that they can consume. But eventually, all sun-loving epiphytes succumb on the ground from a lack of food. And the moment that this happens, the orchid is promptly consumed and decomposed by the mycorrhizal fungus in an eerie form of payback.

Very little rain falls directly onto the forest floor. By the time a raindrop strikes a leaf, it is no longer pure water - it contains minute amounts of dissolved carbon dioxide, dust, smoke particles, and nitrates formed by the action of lightning. Additional nutrients are absorbed from the leaf surface, including nitrogen fixed by cyanobacteria that live in or alongside lichens growing epiphytically on the leaf surface. This 'stemflow' is captured by the aerial gardens of lichens, mosses, and ferns. These, in turn, become saturated with water, which sweeps up insect frass (faeces), bird droppings, and other debris as it trickles and streams along branches down the trees' trunks.

Orchids' roots snake across the rough surface of the bark, seeking out these rivulets of a dilute nutrient soup flowing down the crevices. Besides serving as an anchor, gripping the plant firmly in place, the roots absorb and transport water from their surroundings up to their shoots. To capture every precious drop of water, the roots of epiphytic orchids are covered with a sheath of dead cells called the velamen, which acts like a sponge. When dry, the velamen is silvery white, but when wet it turns green as the chlorophyll within is revealed. Epiphytic orchid roots, therefore, also engage in photosynthesis, augmenting the supply of sugars manufactured by leaves.

The epiphytic lifestyle conveys an ability to exploit a wide range of ecological niches, allowing different orchid species to inhabit different parts of the tree, leading to an enormous diversity of plant architecture. To photosynthesise – to harvest sunshine – epiphytes depend on trees to climb into the light. But life is tough for orchids that choose to live high in the canopy, where they must endure the unrelenting equatorial sun beating down on them day after day. Such plants are adapted to limit water loss. Instead of reducing their leaves to spines, for example (in many respects, these orchids have a lot in common with cacti), they have evolved a thick waxy cuticle. In addition, instead of opening their stomata (the tiny pores in the leaves) in the daytime like most other plants, they open them at night when the air is more humid, allowing carbon dioxide to diffuse into the leaf's interior. The carbon dioxide is then captured within the cells and converted into organic acids that release the carbon dioxide in the daytime, when it can be converted into sugars while the sun shines. This is a special sort of photosynthesis referred to as CAM metabolism. Many species also store water in swollen stems called pseudobulbs. Alternatively, deciduous species, such as lycastes can be epiphytic, lithophytic (growing on rocks), or sometimes terrestrial, shedding their leaves in times of drought.

Orchids adapted to the wet tropics or a constantly humid cloud forest environment have little need for pseudobulbs. The largest group are the New World pleurothallids, which include *Lepanthes*, *Masdevallia*, *Dracula*, *Restrepia*, and *Pleurothallis*. With more than 5,000 species, the pleurothallids make up at least one fifth of

(overleaf)
Vandas have exceptionally long aerial roots that dangle freely in the air. Here they are growing in the Dorothy Chapman Fuqua Conservatory at Atlanta Botanical Garden.
Photo: Philip Seaton

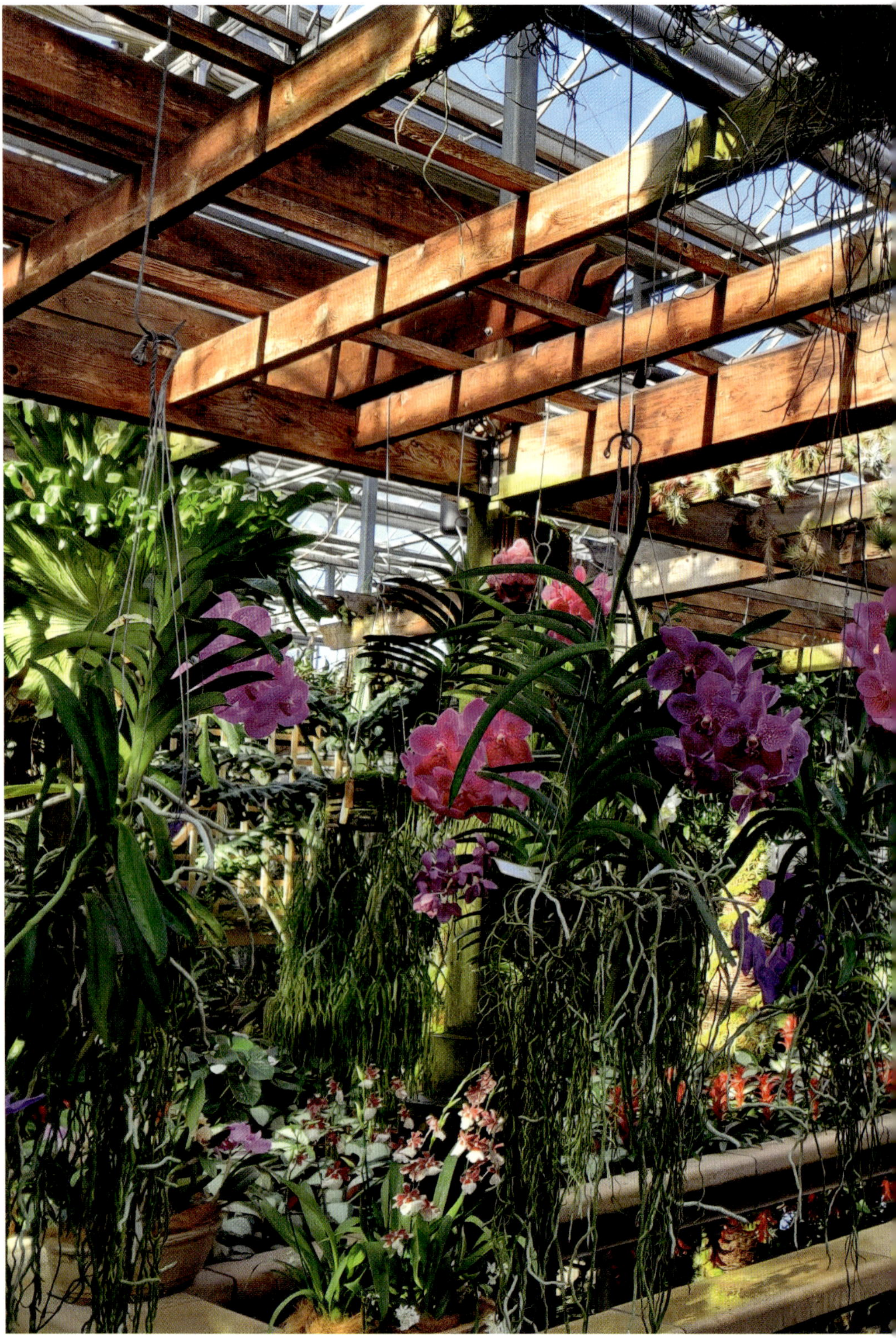

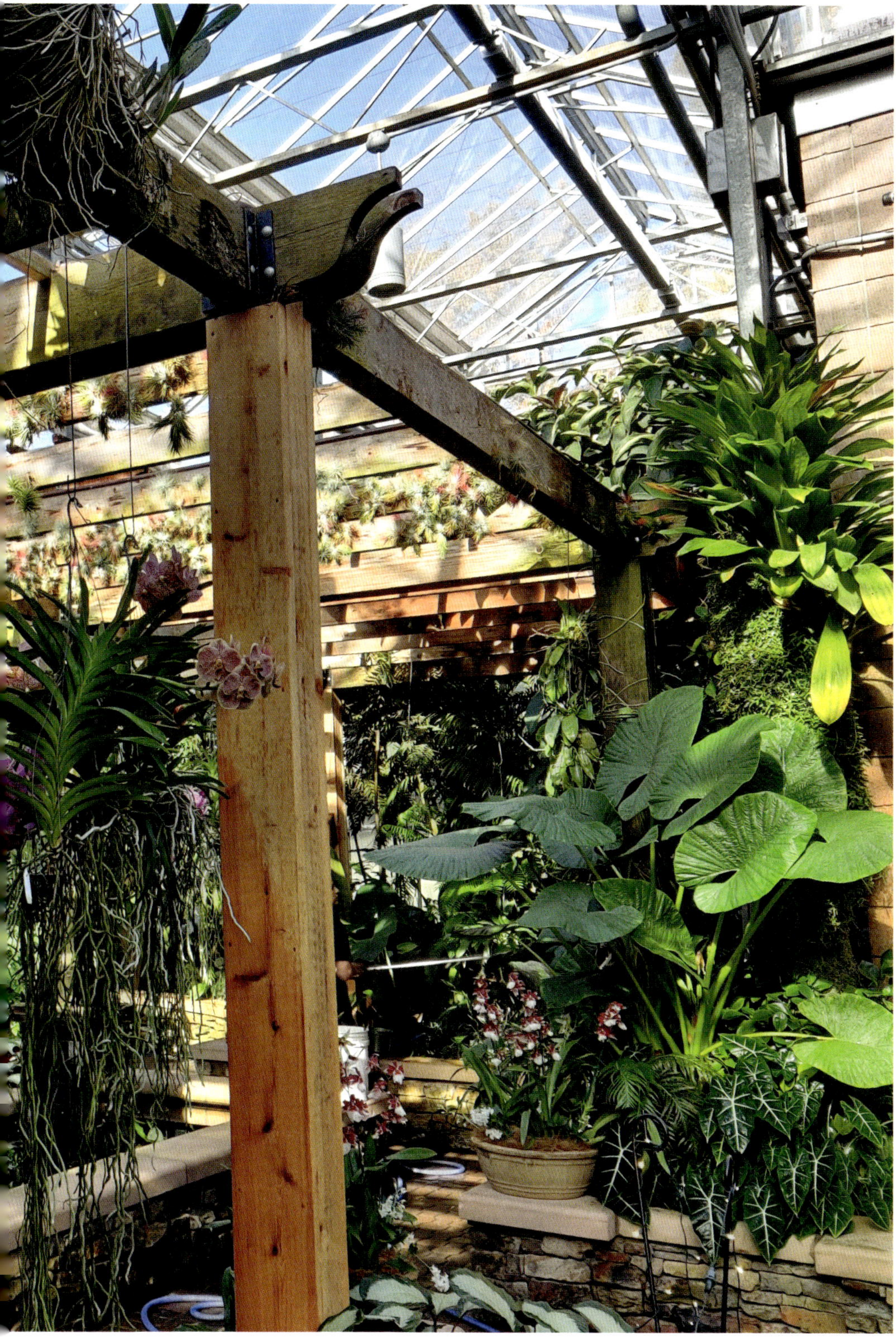

orchid diversity worldwide. Some tropical orchids choose to live fast and die young. Twig epiphytes such as *Ionopsis utricularioides* live a precarious existence at the ends of twigs and are commonly found in abandoned coffee plantations. Similarly, the leafless, diminutive orchid *Dendrophylax porrectus* (syn. *Harrisella porrecta*) is a common inhabitant in citrus groves in the humid Florida air, where it is native but where citrus is not. Thus, epiphytic orchids may share a common lifestyle in terms of the trees they colonise, but they often differ from one another in appearance depending on where in the world they live.

Those that live on the ground

As successful and colourful as epiphytic orchids may be, they are not the only orchids that live in the tropics. There are, in fact, many terrestrial species found on the forest floor that face different challenges. In the tropics, light levels on the forest floor may be just 5% of those high in the canopy above. Nevertheless, tropical terrestrial orchids somehow manage to survive in an underworld filled with decomposition, darkness and decay, feeding on the fungi that blanket the floor and connect to their roots.

By contrast, orchids growing in an English meadow are exposed to full sun. British orchids are typically 'summer green', they become dormant in the autumn and hunker down below ground during the cold winter months. In another contrast, orchids that grow in Mediterranean climates are typically 'winter green', retreating below ground during the hot, dry summer months. Some terrestrials, such as the lady's slippers (*Cypripedium*), are known for their disappearing acts, remaining below ground for several years only to resurface again later, much to the bewilderment of scientists who survey their populations annually. And then there are species such as the bird's-nest orchid (*Neottia nidus-avis*) that have abandoned photosynthesis. These 'obligate mycotrophs' are entirely dependent on fungi as a source of carbon in the form of sugars.

While cracking our Brazil nuts at Christmas, few of us give a thought to their origin – other than that they come from the tropical forests of South America (not just Brazil). Brazil nut trees (*Bertholletia excelsa*) are pollinated by female euglossine bees. Only the females are strong enough to push the hood of the flower aside to reach the nectar. The males play a different role. They pollinate orchids such as the bucket orchid (*Coryanthes vasquezii*), or *Stanhopea* and *Catasetum* species, while gathering a cocktail of fragrant chemicals from the flowers, which they use to attract the females.

After pollination, the flower's ovary swells into a large fruit resembling a coconut, packed with the familiar nuts circularly arranged, like the segments in an orange. Once the fruits fall to the ground, the nuts are dispersed by agoutis, large rodents that are able to chew through the dense, woody fruit's outer wall to reach the nuts. Like squirrels, agoutis scatter and bury some nuts for future consumption, but a few manage to escape, and germinate. The activity taking place on and beneath the Brazil nut tree is an example of a complex web of life involving a myriad of different creatures – all dependent on one another.

Bird's nest orchid (*Neottia nidus-avis*) in Bükk National Park in northern Hungary.
Photo: Philip Seaton

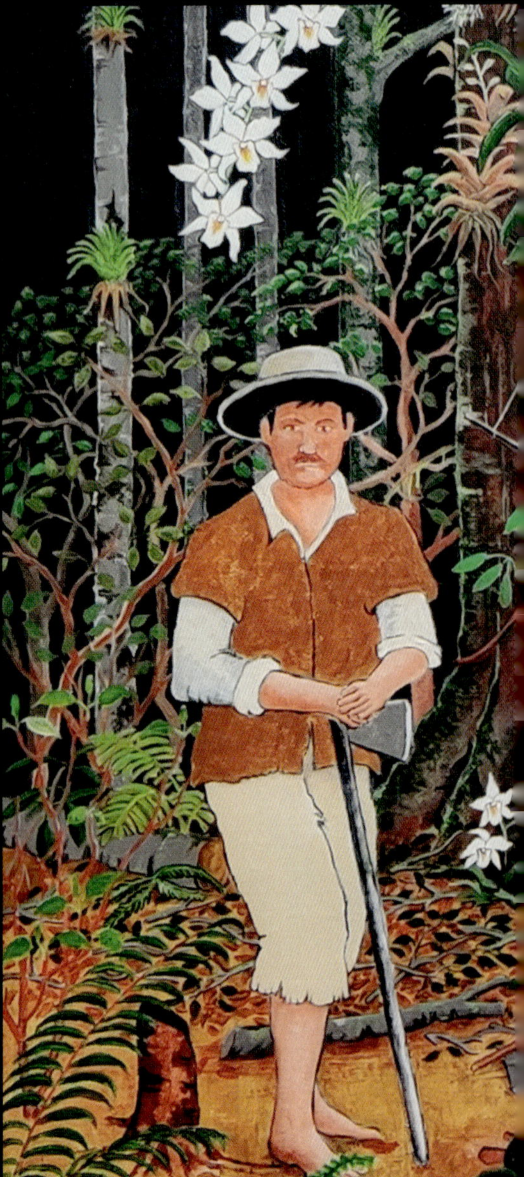

3 **The way we were**

(previous pages)
A painting based on a well-known photograph taken by Albert Millican around 1887, of *campesinos* collecting *Odontoglossum crispum* near Pacho in Colombia.
Painting by Philip Seaton

(right)
Humblot encampment in Madagascar. The photograph (ca. 1885) shows Léon Humblot (with the beard) sitting in the tent and possibly his nephew Henry, on a collecting expedition in Grand Comore. They are seen skinning birds alongside a butterfly net, various jars and collecting tubes and a bag with Humblot's name on it.
Photo from the collection of Johan and Clare Hermans

THE WAY WE WERE

The collectors
The early 19th century was an age of exploration and discovery, of scientific curiosity and a desire to understand the natural world. The tropics were seen as an exotic paradise, with unlimited resources there for the taking. Orchid collection was just one part of a wider interest in collecting specimens. It wasn't only plants that were being sent to Europe. Wilhelm Micholitz, a collector for the nurseryman Frederick Sander, made entomological as well as orchid collections. Léon Humblot shipped not only orchids to Sander from Madagascar, a letter survives in which he mentions sending a live aye-aye (*Daubentonia madagascariensis*) for London Zoo, and a request that Sander should also sell his other lemurs.

Who were those guys?
'Many a collector who set out full of hope to seek his fortune in the form of large quantities of desirable plants, possibly hitherto unknown, never came back, and there was never any news as to how and where he met his end.' Walter Richter, 1965

Arnold, Boxall, Burke, Carder, Chesterton, Cumming, Curtis, Hartweg, Humblot, Johanssen, Lager, Linden, Thomas and William Lobb, Low, Micholitz, Millican, Oversluys, Parish, Pauwels, Roezl, Shuttleworth, Skinner, Wallis, Warszewicz, and Ernest 'Chinese' Wilson. These are the just some of the orchid collectors from the past. Many are commemorated in the names of orchid genera and species. Other would-be collectors perished in pursuit of their dreams and have been forgotten. They were frequently young men of robust constitution looking for adventure, having begun their careers as gardeners, or having been employed in nurseries such as those of Frederick Sander and James Veitch and Sons.

Few left written records of their journeys, but those letters and journals that do survive provide tantalising glimpses of a lost world. A few precious grainy photographs of the activities of these early orchid hunters survive in private collections and in European archives. A faded image, taken by the collector Theodore Pauwels in Colombia in 1896, shows a group of 21 'campesinos' (countrymen) carrying wooden crates on their backs. On close examination, questions begin to form. Where in Colombia was the photo taken? What was in the wooden boxes? Who were the men carrying the boxes? Which village or villages did they come from? Most of the men were barefoot, wearing 'jipijapas', the characteristic straw hats of the region made from thin fibres extracted from the leaves of a local palm-like plant (*Carludovica palmata*). They had been collecting the orchid *Cattleya schroederae*. The photograph must have been taken somewhere in the eastern cordillera of the Andes, in the north of Colombia, where *C. schroederae* is endemic. Photos from what were the early days of

In the early 19th century, the Hackney nurseryman George Loddiges also had a passion for hummingbirds, and instructed his orchid collectors to send him hummingbird skins, eventually amassing an unrivalled collection of more than 200 species.
The endangered marvellous spatuletail (*Loddigesia mirabilis*), a spectacular species with extravagant long tail feathers ending in dark blue rackets, was named in his honour. Today, the marvellous spatuletail still survives in Peru, where it is known locally as '*el colibri mariposa*' (the butterfly hummingbird). Its story reflects the fate of many orchids.

The birds are prized for their feathers, and their hearts are dried into a powder that is sold as a love potion. Their habitat is restricted to a very small area in the cloud forest that is being gradually degraded by agriculture. The remaining populations are very small and declining, and it is listed in Peru as endangered. Hope for its long-term survival probably hinges on ecotourism attracting foreign ornithologists and amateur birders to a visitor centre in the Huembo Reserve. Here, the land is owned by the local community and managed for conservation in partnership with a Peruvian conservation organisation. Profits from tourism are evenly distributed among community members. This story is not unique to Peru and its lovely hummingbird, as birding today is a multi-billion dollar industry. And because most birds fly and many migrate, they need plenty of natural habitats in which to live, breed, and flourish. Orchids are just one beneficiary of the birding industry because orchids and birds often coexist.

photography such as this are rare, but they are powerful reminders of how the early collectors lived and worked.

A dangerous game
'Those parts of Madagascar which especially attract botanists must be death-traps indeed! Mr Léon Humblot tells how he dined at Tamatave with his brother and six companions, exploring the country with various scientific aims. Within twelve months he was the only survivor.' Frederick Boyle, 1893

It was said that if you fell in the Magdalena River in Colombia, if you didn't drown, you were in danger of being eaten by caimans, and if you went ashore, you could be bitten by venomous snakes. The explorers Alexander von Humboldt and Aimé Bonpland complained bitterly about the mosquitoes. These insects weren't just a nuisance – many early travellers succumbed to tropical diseases carried by them. Both yellow fever and malaria appear to have been introduced into South America with the transatlantic slave trade. The American collector Robert Grey, describes travelling to Colombia with his brother, who soon died of yellow fever in Honda. Likewise, George Ure Skinner succumbed to this dreadful haemorrhagic fever. It is thought to have arrived from Africa some time before 1650, accompanied by its mosquito vector, *Aedes aegypti*, which would have bred in the casks that provided drinking water. The virus became endemic in the monkey populations in the forest canopy, where it was transmitted by a native mosquito, *Haemagogus spegazzini*. It could be transmitted to humans whenever trees were felled, the woodcutter returning home while incubating the disease, and becoming the source of an urban outbreak.

Today's researchers are not entirely immune to tropical diseases and other dangers, and rarely do they conduct fieldwork alone. When working with orchids in the Reserva Ecológica del

Pedregal de San Ángel at the University of Mexico in Mexico City, researchers are obliged to take a medical kit with them containing snake antivenin because of the risks posed by rattlesnakes. In Australia, the dangers posed by wildlife tend to be overstated, but they do exist. Researcher Zöe Smith relates treading on a brown snake (*Pseudonaja textilis*), one of the world's most dangerous reptiles. Unless antivenin is administered promptly, death usually ensues within a few hours of being bitten. Happily, she stood on its tail rather than its head, and it slithered away rapidly. Solitary kangaroos can be equally dangerous. And it's not just the wildlife. Parts of Latin America remain out of bounds to researchers because of the continued presence of *narcotraficantes* - drug traffickers.

The scale of collection
'I place Colombia first because I believe its flora has no equal in the world, and taking the orchids especially I doubt very much if another region can be found where such great numbers of species occur.'
John Lager, 1907

It is difficult to grasp the past abundance of some orchids in their natural habitats, and the scale of their removal from tropical forests during the late 19th and early 20th centuries. Reading accounts of the extraction of hundreds of thousands of plants from the wild, we may wonder if this isn't an exaggeration. Yet, by 1886, tens of thousands of the 'Pacho' form of *Odontoglossum crispum* were being imported from Colombia by Frederick Sander into his nursery in St Albans. Frederick Boyle describes 12 glasshouses set side by side, each 55 m (180 ft) in length. In the first, there were no fewer than 22,000 pots containing *O. crispum*. In the same year, Frederick Burbidge stated that *O. crispum* is 'being grown literally by the million in this country'. In 1878, William Bull of Chelsea proclaimed that their nursery had received from Colombia around two million plants, mainly *O. crispum* and *Cattleya mendelii*. *Miltoniopsis vexillaria* was first seen by the German collector, Gustav Wallis, in the collection of architect Jean Lalinde when he arrived in Medellín in 1866. A little more than 20 years later, a receipt from Cartagena de las Indias, dated 6th July 1889, shows that Albert Millican paid a customs duty of 18 pesos oro for the export of 125,000 'Josephinas' (*M. vexillaria*) destined for Chelsea, London. An entry from Theodore Pauwels' Medellín diary of 26 May, 1896 lists 56 cases of orchids, containing 3,570 *Cattleya warscewiczii*, 2,040 *C. aurea*, 538 *Oncidium kramerianum* (syn. *Psychopsis krameriana*), and 100 *Miltoniopsis roezlii*.

By 1887, the orchid trade in Colombia was of such significance that it began to appear in the *Diplomatic and Consular Reports on Trade and Finance*, alongside figures for more familiar exports such as bananas and coffee. Collectors were obliged to pay customs duty. Report No. 456 tells us that in 1887, Colombia exported

(97.6 imperial tons) of 'plants (orchids, etc.).' In 1899, the government of Colombia issued Decree 473, which forbade the cutting down of trees. Although aimed specifically at preventing the destruction of trees to obtain rubber, it meant that everyone who wished to exploit the 'National Forests' needed to obtain a licence and to provide an assurance that trees would not be cut down or destroyed to harvest any natural product.

The plants
Many plants were imported into Europe as enormous specimens. Boyle tells us that a piece of *Cattleya mossiae* sent home by the collector 'Mr Arnold' had been described as, 'enclosing two great branches of a tree, rising from the fork below which it was sawn off – a bristling mass four feet thick and five feet high; two feet more must be added if we reckon the leaves. As for the number of flower scapes it bore last season, to count them would have been the work of hours; roughly I estimated a thousand, bearing not less than three blooms, each six inches across . . . I doubt not that the forest would be scented for a hundred yards round.'

Of the millions of orchids that were imported from around the world, no species was considered more beautiful or desirable than *Odontoglossum crispum*. James Bateman provides us with a clue to this craze when he writes that a plant discovered in 1863 by John Weir, 'in the gloomy forests that clothe the slopes of the lofty

Painting of *Odontoglossum bluntii* (*Odontoglossum crispum*) in John Day's Scrapbooks.
Painting: RBG Kew

mountain-ranges at the rear of the City of Santa Fé de Bogotá' was 'not only a new species, but one of surpassing beauty'. Bateman named the plant *Odontoglossum alexandrae* (how and why orchid species may have their names changed is explained later, under 'What's in a name?', page 96). John Day notes on his illustration of a fine white form of crispum (labelled *Odontoglossum bluntii*) that he had painted in 1866 that, 'The texture of the flower is exquisitely beautiful – it is dazzling white with the light sparkling through each of its countless cells, as if made of hoar frost.' Day recorded many of the early orchid importations in his 'Scrapbooks'. Clearly entranced by *O. crispum*, he painted different specimens on over 40 different occasions between 1865 and 1887. Not only are his watercolours depicting their flowers of historical interest, but the accompanying notes are equally revealing. Day took a lot of time and trouble to measure and produce accurate representations. On the page illustrating *O. crispum* var. *veitchianum* he tells us that he has measured the flowers with compasses and taken care to represent the colour and number of spots accurately. In another instance he apologises for the lack of accuracy of the purple colouration, because he had done the painting by gaslight.

Decline
Towards the end of the 19th century it was becoming clear that many orchids were disappearing from their natural habitats, and could no longer be found in London's sale rooms. As an example, the great attraction at a sale at Messrs. Protheroe and Morris's Rooms in 1894 was the 'fine importation of the beautiful *Cattleya dowiana aurea*, which unfortunately, is generally supposed to be now becoming rare in its native habitat.'

In Colombia there was some hope of a respite from collectors when an 1887 report in *The Gardeners' Chronicle* informed growers that civil war was likely to prevent further explorations for some time. Nevertheless, around 1896, Carl Johannsen was able to write to Frederick Sander from Colombia: 'I shall despatch tomorrow thirty boxes, twelve of which contain the finest of all the aureas (*Cattleya aurea*), the Monte Coromee form, and eighteen cases contain the grand Sanderiana type (*Cattleya warscewiczii* var. *sanderiana*), all collected from the spot where these grow mixed, and I shall clear them all out. They are now nearly extinguished in this spot, and this will surely be the last season. I have finished all along the Rio Dagua, where there are no plants left; the last days I remained in that spot the people brought in two or three plants a day and some came back without a single plant.'

A first edition of Albert Millican's book was recently sold by Sotheby's that contained a collection of 13 letters, including three signed by Millican himself, relating to plant hunting. In one he wrote: 'I have been 2 months seeking Mendelli [*Cattleya mendelii*] and have only got 14 cases – it is almost impossible to find plants in all the old districts, they have been taken.' In 1889 *Odontoglossum crispum* had become extremely rare in the immediate vicinity of Pacho and, to procure it, it was necessary for the collectors to penetrate further into the mountains and search around the other villages at higher altitudes. Instead of returning every night with their spoils, the collectors were obliged to carry provisions for a week. In 1901, Boyle wrote: 'The time is very close when *Odontoglossum crispum* . . . will arrive by tens and units instead of myriads – and then will arrive not at all.' By 1907, Florent Claes described a much-changed environment. Most of the forests surrounding Pacho had been cut down to make room for pasture lands or for maize or tobacco plantations.

Collateral damage
Tens of thousands of trees were either cut down or burned by collectors in Colombia. Wilhelm Hennis relates that, having arrived in Frontino, he set off in search of *Cattleya aurea*. After three weeks' search, he found that the majority were growing on trees 'similar to *Tamaranda indica*, the wood of which was so hard that the axes just flew off it, but it was rich in inflammable resin. We hacked pieces out of these trees, carried a pile of dead wood up into them and in this manner set the tree on fire. It burned slowly until it fell, this usually taking two or three days. It was always an experience when such a gigantic tree fell down with a thunderous noise, which echoed round the mountains, particularly when this happened during the night.' He was disappointed that he managed to gather just five crates of *C. aurea*.

One image in Albert Millican's *Travels and Adventures of an Orchid Hunter*, is framed on one side by a man leaning on his axe, with Millican himself on the other side, partly obscured by the smoky haze of the campfire. A group of men, mostly barefoot, are seen sitting on ground cleared of trees. They are eating from bowls and have their axes by their sides. In the centre of the image, a rifle can be seen hanging from a large tree, below which are a large number of *Odontoglossum crispum* heaped up against the trunk, some in flower.

Writing in the *Transactions of the Massachusetts Horticultural Society* in 1907, John Lager paints a vivid picture: 'Men are sent out in every direction, covering as large a territory as possible, in order to secure the plants in the least space of time. As a rule two or three men band themselves together, and take provisions enough to last a week or more. They are armed with the indispensable machete, and axes occasionally, also with shotguns, and carry string bags in which to bring out the plants. These men bury themselves in the woods, and when sighted the trees are usually cut down, unless the plants are found quite low down. To climb the trees is no easy task in the tropical forests where there is a network of climbers and other vegetation around the trunks. There is also the possibility of

(top)
Collecting *Cattleya schroederae* in Colombia at the end of the 19th century.
Photo: from the collection of Th. Pauwels, Roger Bonte and Chris Loncke

(above)
Staff at Moore Ltd orchid nursery in 1909, unpacking a delivery of vandas from Rangoon in Burma (now Yangon in Myanmar). Later that year, two of the employees, Mansell and Hatcher, took over and renamed the company.
Photo: Chris Barker,

poisonous insects, snakes and scorpions being hidden in the plants; hence few men will undertake the climb; but, whatever mode is chosen, the plants are loosened from the trunks or branches by running the *machete* under the plants; when once loose they are tied together with some kind of string, put in the bag, and the march continued until another plant is found. Towards night the men return to their camp, where the plants are spread out beneath some tree to protect them from the sun.' The process of collecting was often very wasteful. Hennis tells us that, in the winter of 1882–1883, he sent around 200 crates of *Cattleya trianae* from the state of Tolima, having thrown away three times as many plants that had either been damaged in transit or during the felling of the trees on which they grew.

Having collected the orchids, the problem was then how to get the booty out of the cloud forest. The challenge was particularly acute with plants collected from the cooler regions, such as *Odontoglossum crispum*. Hennis writes: 'Owing to the peculiarly soft nature of the plants, they are so liable to decay that in some instances seven-eighths of the consignment would be dead on arrival.' The locally-employed collectors carried the plants on their backs to the edge of the woods, where they were cleaned and prepared for packing in spacious baskets (constructed from thin poles cut in the forest) transported on the backs of oxen to Pacho - a journey of five days. Here the orchids were placed in solid wooden cases, which were sent down to the banks of the Río Magdalena on the backs of mules, for transfer to river steamers that took them to the coast. Upon reaching the river port of Barranquilla five to seven days later, the orchids could languish for weeks in the hot and steamy atmosphere, before being loaded on board a sailing ship or steamer. 'Freight services at that time were irregular and not bound to any time table,' notes Hennis. 'The crates were transported over the sea in damp, dark and warm holds, without the light, air and freedom essential to the needs of living plants.' Not surprisingly, tens of thousands of orchids perished before they reached English shores and, having arrived, many did not recover from the effects of the journey. Indeed, it can seem little short of miraculous that any orchids reached England alive.

Timing was everything. In September 1890, Oversluys wrote to Frederick Sander from Cobán in Guatemala to say that he had collected many fine plants, which he had cleaned and placed under the shade of trees. Unfortunately, the wet season had arrived, with heavy rainfall. But he was hoping to keep the orchids in good condition until they could be taken away on the backs of men as there were no mules available. Nurserymen and growers preferred to receive plants in the spring, when they would have the whole summer in front of them, and the best chance of becoming established. Time of year and weather were also critical for the orchids' survival during their journey to Europe or North America.

Ideally plants were exported during the dry season, when they were dormant. If they were still active, or just beginning to become so, any new growth would be weak and spindly due to the lack of light, and the plants were liable to die soon after being unpacked. B. S. Williams says: 'I have seen many a fine mass of Cattleya with all the leading growths completely rotten. This is sad to contemplate, involving as it does the extermination of the plants in their native homes, and loss of time to the collector, which, combined with loss of money, vexation, and disappointment to the cultivator at home, has a most depressing influence on orchid growing.'

In the early 19th century, the nurseryman George Loddiges complained that only one plant out of 20 would survive the journey to Europe. The situation didn't improve. Williams reported in 1877 that the principal cause was, 'the rapacious appetite of the collector – the boxes being overcrowded by his sending home thousands instead of being satisfied with a few dozens, and in consequence all arrive dead.'

The invention of the Wardian case in the early 19th century – essentially a small, airtight transportable greenhouse - had the potential to dramatically reduce the losses. Although we can find no evidence that they were used to export plants from Colombia, for example, certainly they were used to some extent, as Williams writes: 'Cases in which orchids are sent home ought to be made strong and roofed with good stout glass not easily broken; for I have often seen plants spoiled by the glass being fractured. Through an accident of this kind, salt water and cold air get in, both of which are very injurious. All cases should be air and water-tight – the case must not be placed during the journey near heated surfaces or fires in the vessel. I have seen many boxes of plants spoiled by being set in such positions, the leaves being completely dried up; they ought to be placed in a moderately warm situation, but by no means near any fires.'

Cultivation

In 1874, James Bateman wrote that despite repeated warnings from collectors in the field, orchids 'invariably succumbed under the stifling atmosphere [in stove houses] to which they were remorselessly consigned.' Once their needs were understood, cultural techniques soon improved, and large private collections were established. In the 1890s Joseph Chamberlain's grower, Burberry, wrote extensively about orchid culture, emphasising the importance of mimicking the natural conditions under which the orchids were found. Plants were generally grown in terracotta pots or wooden baskets, and washing the pots was a constant task. Williams recommended using, 'good rough fibrous peat and live sphagnum moss.' There was a lot of experimentation. Osmunda fibre, derived from the densely matted roots of the royal fern (*Osmunda regalis*), became the potting medium of choice, providing ample aeration around the roots, while being water-retentive. It only decomposed slowly, and provided a source of nutrients as it did so.

Growing orchids was (and still is) a labour-intensive operation. Greenhouses were heated by cast-iron hot water pipes which produce a buoyant atmosphere with gentle air circulation and a high relative humidity. Gardeners and their apprentices were responsible for stoking the boilers throughout the night, often sleeping in the boiler house. By day, they opened and closed greenhouse vents. Shading was provided by wooden lath blinds that were continually deployed or rolled up in the daytime to maintain optimum light levels, while maintaining stable temperatures during the summertime. Regular damping down in hot weather maintained the necessary high relative humidity.

The nurseries and the nurserymen
Loddiges and Sons began cultivating orchids in Hackney, London, in 1812. It was the principal (and first) orchid nursery in England, and arguably in Europe. Later came the enormous nurseries of Veitch in Chelsea and Exeter, and Sander in St Albans and subsequently in Bruges, Belgium. Frederick Boyle describes a visit in 1891 to Sander's vast premises at St Albans, which was then the world's largest orchid nursery, served by its own railway branch line. In the 'Importing Room', Boyle says, 'cases are received by fifties and hundreds, week by week, from every quarter of the world, unpacked, and their contents stored until space is made for them up above. Orchids everywhere! They hang in dense bunches from the roof. They lie a foot thick on every board, and two feet thick below. They are suspended on the walls. Men pass incessantly along the gangways, carrying a load that would fill a barrow. And all the while fresh stores are accumulating.' The unpacking was undertaken carefully, not only to avoid damage to the plants and save the valuable packing material, but also because the packing cases sometimes contained unwanted stowaways, such as rats, scorpions, large venomous centipedes, stinging ants, and dangerous snakes.

Collectors and nurserymen would often send herbarium specimens to taxonomists such as John Lindley, Sir William Hooker at Kew, and Heinrich Gustav Reichenbach in Germany. Such an arrangement was mutually beneficial: the taxonomists would receive new material for identification, and the nurserymen would have their plants scientifically identified and named. Doubtless it was in the commercial interests of nurserymen to give each small difference in the flowers a distinct varietal name: they were 'splitters'.

The private growers
In the late 19th and early 20th centuries, orchid growing was primarily a hobby of wealthy 'gentlemen growers'. It provided a way of demonstrating that they were persons of culture and good

taste. We can think of them as belonging to an exclusive club, where members were engaged in friendly rivalry. Often they could be seen in the London sale rooms, bidding against one another. Writing in 1886, Castle gives us an idea of some of the eye-watering prices paid. In 1856, the Duke of Devonshire bought a *Phalaenopsis amabilis* for £68 5s, 'the fine specimen that Fortune purchased in the Island of Luzon for a dollar, ten or twelve years previously.' He writes that Mr Fairrie of Liverpool bought a *Laelia superbiens*, said to be the finest specimen in Europe, with 220 pseudobulbs and measuring 17 feet in circumference, for £36 15s. He adds: 'At a sale in Liverpool, *Laelia anceps* var. *dawsonii* was sold for £46.' John Day's collection was sold for an amazing £7,059 in 1881.

The sizes of the collections were staggering. Castle says: 'Writing in 1841, Bateman said that the collections of orchids were innumerable,' and he wonders, if that was the case then, what would be said now? He continues: 'The plants were then numbered by hundreds now amateurs possess thousands, and one, Mr. R. Warner, has even had as many as 12,000 plants of one species, *Odontoglossum crispum*.' Sir Jeremiah Colman, who established Colman's Mustard, grew 30,000 species at Gatton Park (and 25,000 hybrids). Joseph Chamberlain had two fresh orchid corsages sent to him from Birmingham to wear in his buttonhole each time he attended Parliament. Chamberlain's greenhouses have gone, but it is still possible to visit his Venetian Gothic style home, Highbury, in Birmingham. To one side of the magnificent and richly decorated two-storey hall is the drawing room, with a ceiling inlaid with panels of satinwood and walnut. At the far end is a colonnade of alabaster columns and a door inlaid with panels of walnut and sycamore. Standing before the door one can imagine walking through the palm house and the fernery that lay beyond, leading in turn to a corridor 61 m (200 ft) long that, on its south side, provided access to 13 span greenhouses in succession. The 'Mexican House' alone contained a collection of one thousand *Laelia anceps*.

Wealthy amateur growers, such as Lionel Rothschild and Baron Bruno Schröder, had their own laboratories. At that time, health and safety considerations were a distant dream, and toxic chemicals were routinely employed as sterilising agents. Rothchild's laboratory had an outer chamber, or 'sterilising room', and an inner chamber, 'the sowing room'. Before use, both were filled with phenol (carbolic) fumes to cleanse them. Only the operator was allowed to enter the rooms and, to avoid contamination, had to wash his or her hands with 'NEKO' germicidal soap, containing 2% mercuric iodide. The packaging warns: 'Certain individuals are extremely sensitive to mercury compounds'! Later in the 20th century, amateur growers were still being advised to soak cotton-wool plugs for flasks in a solution of picric acid and mercury (II) chloride, the cotton wool turning yellow before being dried in trays in the sun. By the 1970s, however,

Miltoniopsis vexillaria was collected by Albert Millican in Colombia in the 19th century, and sent to growers in England.
Photo: Luis Eduardo Mejía

cotton-wool plugs were largely replaced by rubber stoppers incorporating a glass tube with a small cotton wool plug as a 'breather'.

What happened to the millions of orchids that were imported into the UK, Europe, and the USA? The double whammy of two world wars played a pivotal role in the demise of many of the large private orchid collections and nurseries in Europe. Some survived, but most did not. Shortages of fuel to heat the greenhouses, bombing raids, and the loss of staff to serve in the armed forces all played a part. The nursery of Pauwels in Belgium was completely destroyed by shelling in 1918. Today, many of the big houses of orchid collectors remain, but the greenhouses are long gone. Against the odds, perhaps, viable populations of *Odontoglossum crispum* can still be found in its native Colombia.

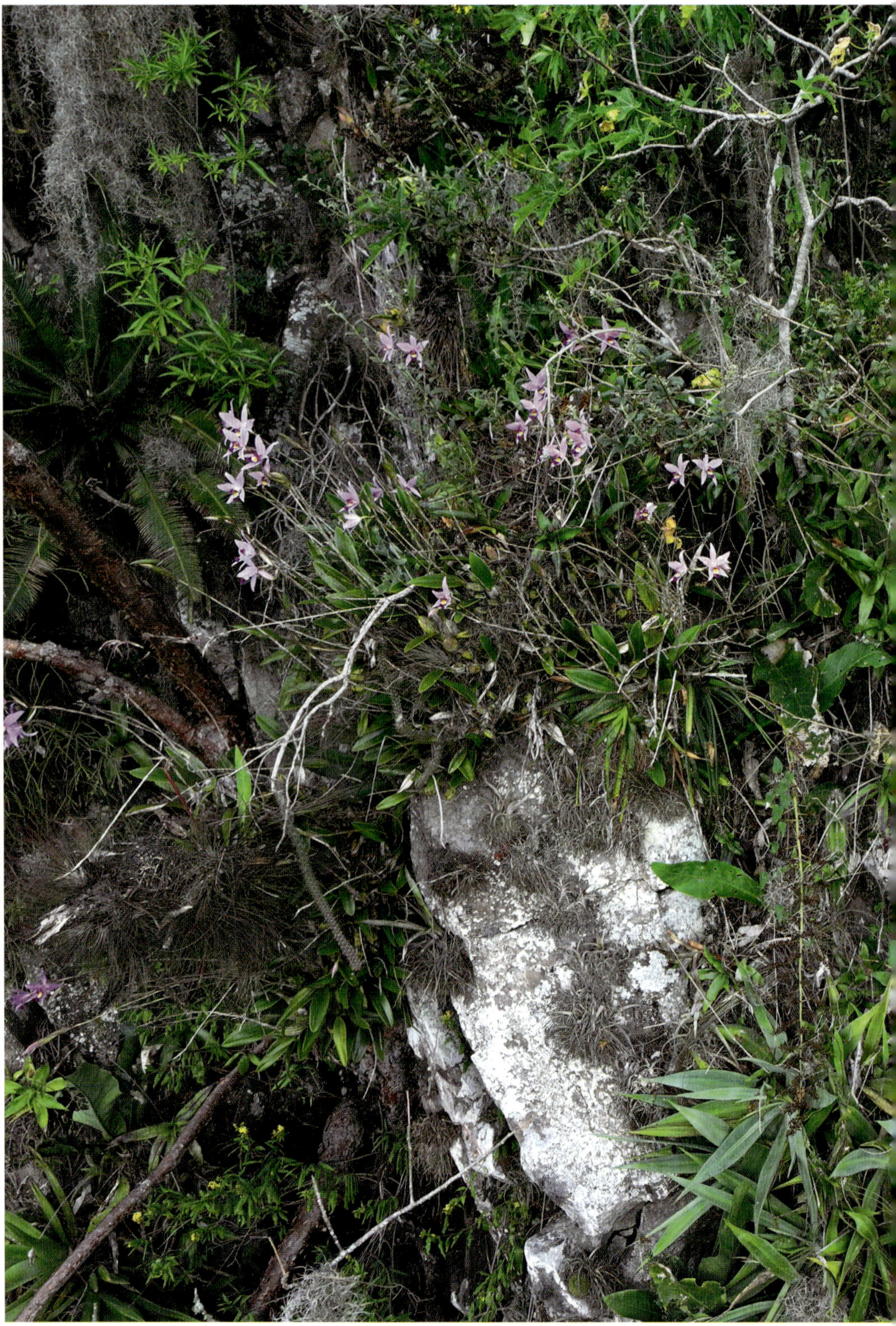

4 Orchids (almost) everywhere

Apart from Antarctica, orchids are found on every continent, and comprise about 10% of the world's flowering plant species. Some genera and species are circumpolar or pan-tropical in their distribution, whereas others are confined to very specific locations. Here we are only able to scratch the surface, describing some of the habitats with which we have some familiarity, such as those in the Americas and, to a lesser extent, Europe, Africa, Asia and Australia. The fate of the world's orchids is inextricably linked to the preservation of their diverse natural habitats. With limited resources, difficult decisions will have to be made.

(previous pages)
Laelia anceps growing on a roadside cliff in Alto Lucero Veracruz, Mexico.
Photo: Andrés Ramos

'The day has passed delightfully. Delight itself, however, is a weak term to express the feelings of a naturalist who, for the first time, has wandered by himself in a Brazilian forest. The elegance of the grasses, the novelty of the parasitical plants, the beauty of the flowers, the glossy green of the foliage, but above all the general luxuriance of the vegetation, filled me with admiration. A most paradoxical mixture of sound and silence pervades the shady parts of the wood. The noise of the insects is so loud, that it may be heard even in a vessel anchored several hundred yards from the shore; yet within the recesses of the forest a universal silence appears to reign. To a person fond of natural history, such a day as this brings with it a deeper pleasure than he can ever hope to experience again. After wandering about for some hours, I returned to the landing-place; but, before reaching it, I was overtaken by a tropical storm. I tried to find shelter under a tree, which was so thick that it would never have been penetrated by common English rain; but here, in a couple of minutes, a little torrent flowed down the trunk.' Charles Darwin, 1839

A first visit to the tropics often comes as a surprise for people accustomed to living in temperate climates. Upon stepping off the plane, the traveller is enveloped in a deliciously warm, humid blanket, with the sweet perfume of the earth. Often thought of as being unremittingly hot, the reality is that the tropics are bathed in a narrow (warm) temperature range year-round, unlike temperate regions, which are marked by seasonal extremes in temperature. Such conditions promote nearly continuous plant growth, and it is therefore no surprise that the tropics are home to the greatest diversity of species, since natural selection and evolution are in motion 365 days a year without interruption.

Europeans first explored tropical forests when Alexander the Great crossed into the Punjab in India in 327 BCE. The term 'jungle' is derived from the Hindi *jangal*, meaning a dense and impenetrable forest. But the term 'tropical rainforest' was coined more recently, in 1898, by the German botanist A. F. W. Schimper. He classified all of the forests in the permanently wet tropics as *tropische regenwald* and recognised four different types based on their woody vegetation: rainforest, savannah forest, monsoon forest, and thorn forest. The reality is much more complex and interesting. Not all tropical forests are equal – some are ever-wet, while others are seasonally dry. Tropical forests on different continents look, feel, and sound different. Entering a tropical forest in the Americas, for example, the visitor will encounter bromeliads cloaking the branches of the trees, alongside hummingbirds or toucans, and New World monkeys with their prehensile tails. The lowland forests of South-East Asia are notable for their tall, clear-trunked dipterocarp trees, carnivorous pitcher plants, parasitic rafflesias, gibbons, and orang-utans. Madagascan forests may be alive with lemur species. The same is equally true for orchids. The large and showy cattleyas and brilliantly coloured masdevallias are endemic to Central and South America, whereas dendrobiums reside in South-East Asia and Australia. In Madagascar, a large share of orchids are variations on the same theme of white, starry flowers with long nectar spurs.

Cloud forests, and dry tropical forests, are the most threatened of the major tropical forest types in many parts of the world. Cloud forests are a rare and fragile ecosystem, making up no more than 2.5% of the world's tropical forests. Periodically enveloped in clouds, they are home to the greatest orchid diversity. Walking through a cloud forest is a deliciously cool experience. The silence is only broken by the dripping of water on leaves. The atmosphere is humid, and it is constantly wet underfoot. The trees are festooned with aerial gardens of orchids, lichens, mosses, ferns and, in the Americas, tank bromeliads which help to maintain a humid environment for other plants by storing water in their rosettes.

Then again, we shouldn't be talking about 'the cloud forest', but about cloud forests. They are all different. When thinking of orchids growing in tropical biomes, coniferous trees don't often spring to mind, and yet evergreen cloud forests with huge pines that support a rich orchid flora are found in Mexico, and the cloud forests in Taiwan are rich in other types of conifers. 'Stalky' Dunsterville aptly said that in Venezuela, 'cloud forests occur like plums in a plum cake'. They are islands in the sky, with individual mountain tops serving as centres of biodiversity in their own right, home to considerable numbers of endemic species.

Tropical dry forests, on the other hand, are found in regions marked by a prolonged dry season lasting several months, with most rain falling during a (usually) brief wet season. Such forests are typically found in very warm regions in the tropics, where the mean annual temperature is greater than 17°C (63°F), and where rainfall is in the range of 250–2,000 mm per year (10–80 in). The orchids growing there are tough, typically possessing thick, leathery leaves and pseudobulbs for storing water during hard times, exemplified by *Cattleya trianae* in Colombia, and *C. maxima* in southern Ecuador and northern Peru.

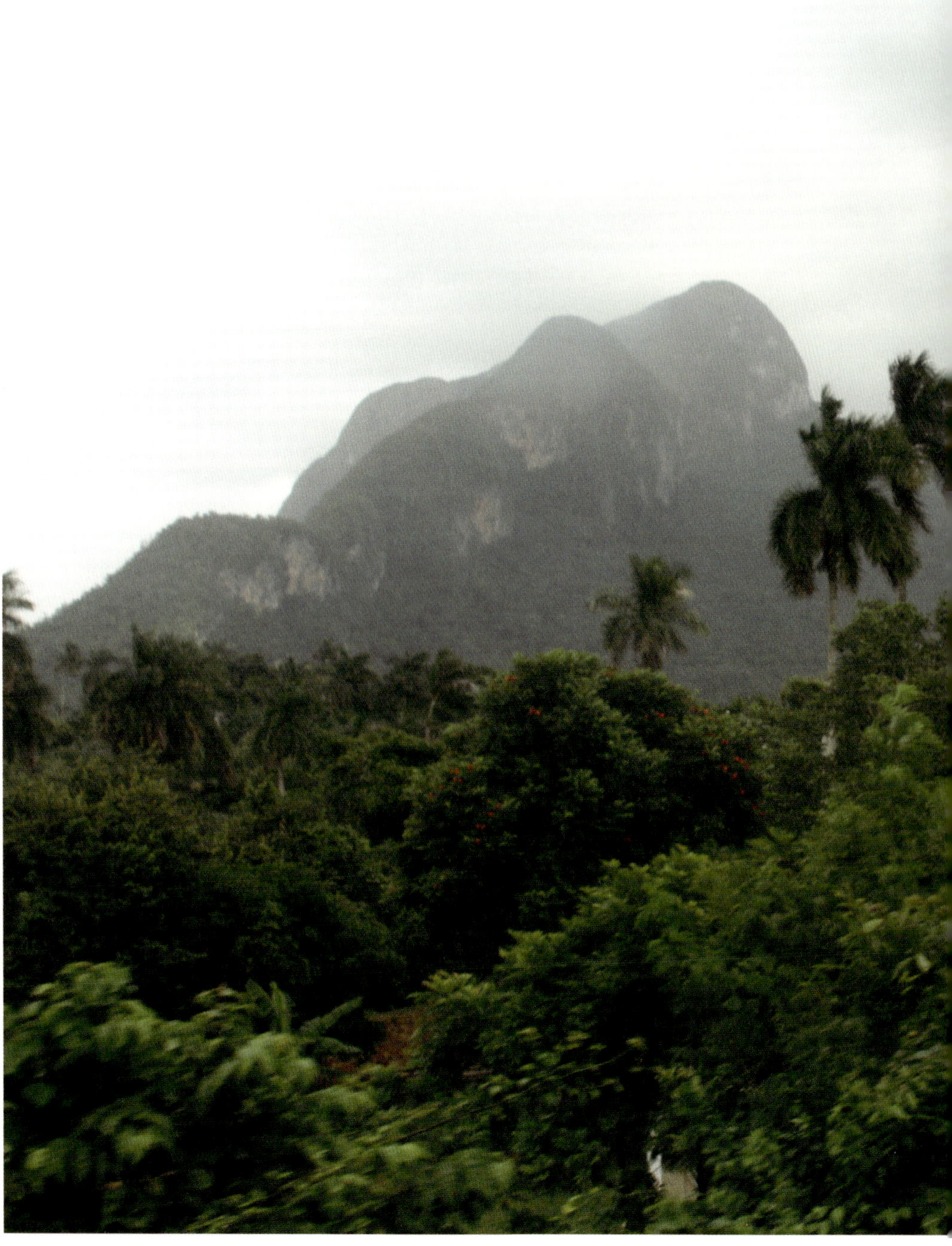

5 **Islands**

Because of their isolation, islands often harbour a unique flora and fauna shaped by the forces of evolution. In Indonesia, for instance, a small group of monitor lizards, probably originally from Australia, colonised a cluster of islands in the Sunda group around four million years ago and quickly evolved into the largest and heaviest lizard in the world – the Komodo dragon, *Varanus komodoensis*. On a much smaller scale, the ancestors of a group of inchworm caterpillars in the genus *Eupithecia* may have arrived in the Hawaiian archipelago on driftwood as hungry stowaways. Over time and out of necessity, these caterpillars abandoned their vegetarian diet of leaves and adapted to catching and eating insect prey, evolving into the world's only known carnivorous caterpillar species.

The genesis of our understanding of island biogeography began in 1799, when Alexander von Humboldt sailed west to the Canary Islands, and onward to South America in a small Spanish ship that somehow eluded a British naval blockade. The purpose of his journey was . . . 'to find out how the forces of nature interact upon one another and how the geographic environment influences plant and animal life'. What von Humboldt revealed was that mountains also harbour unique communities or patterns of life forms. Essentially, for every 1,000 m increase in elevation, there is a 9.8°C (17.6°F) drop in temperature. For this reason, cold-sensitive epiphytic orchids are nowhere to be found on mountain peaks above 4,000 m (13,000 ft), even on the equator itself.

Thus, there are islands surrounded by water that serve as ecological outposts for many different life forms, and islands on continents (mountains) that ascend into the sky providing a home for cooler-adapted species. Everywhere you look, whether on remote islands or the tallest mountains, life forms are endlessly searching for the good life – food, a place to live, and a place to reproduce. Most never find that niche and perish without a trace. Others are much more fortunate. But life on an island paradise has a downside. There is no place to hide from new arrivals that compete for the same limited resources, or from the diseases they carry from the outside world. This is why species endemic to islands are particularly vulnerable to extinction. Perhaps no place illustrates this concept better than the Hawaiian archipelago in the Pacific, where many endemic bird species have succumbed to avian malaria spread by invasive mosquitoes that were inadvertently introduced by humans.

Hawaii: three lonely orchids in paradise
Few places on Earth are more remote than the Hawaiian archipelago in the Pacific, roughly 3,000 km (1,850 mi) from the North American continent. Formed by an ongoing volcanic ooze originating beneath a weak spot in the Pacific Plate, the Hawaiian Islands are geological newcomers that have a lifespan of about

(previous pages)
Many of the mountains of the island of Cuba, such as the Sierra de los Órganos, are probably home to yet-to-be-discovered orchid species.
Photo: Philip Seaton

6 million years. The youngest and largest is Hawaii, often referred to as The Big Island. It emerged 0.7 million years ago and rises over 4,000 m (13,000 ft) through the volcanic portal known as Mauna Kea. To the north-west lies an elder sibling, Kauai, which is nearing the end of its life as it slowly slips into the ocean, beaten down by years of endless erosion, carrying its cargo of plants and animals to a watery grave.

Had Darwin sailed north beyond the equatorial Galápagos Islands to the Hawaiian archipelago, he would have studied colourful birds with strange beaks, such as the ʻiʻiwi and ʻakiapōlāʻau, instead of finches. But he would not have seen tortoises or marine iguanas, because Hawaii is far too remote for even reptiles to survive as stowaways from a nearby continent on pieces of driftwood. If he were lucky, he would have spotted one of the three native orchids, probably *Anoectochilus sandvicensis*, which the locals refer to as Ke kino o kanaloa. All three orchid species are terrestrial and not as showy as one might expect in such a picturesque place as Hawaii.

Compared to the Galápagos Islands, Hawaii is a far lonelier place with fewer species. But species that did arrive there, whether by driftwood or through the air, found a new world rich in potential ecological niches to exploit. One by one, each new arrival began to evolve and co-evolve into the unique flora and fauna found there today. Among the flowering plants, about a thousand species are native to the islands, and around 89% of these are endemic (found nowhere else).

 Orchids are not alone in taking a back seat to their cousins on the mainland in terms of abundance. Butterflies, for example, rival orchids in their global diversity (20,000 species compared to 30,000-plus orchids), yet there are only two species found naturally in Hawaii – Blackburn's bluet (*Udara blackburni*) and the Kamehameha butterfly (*Vanessa tameamea*). According to the entomologists Francis Howarth and William Mull, it takes roughly 175,000 years for a new insect species to become established on the archipelago from a distant landmass.

What makes butterfly and orchid species so sparse on islands? Both have highly specialised diets, especially when they are very young and most vulnerable. Caterpillars often feed on a select group of plants, whereas orchid protocorms feed on a select group of fungi. Without food sources, these young 'larval stages' succumb to starvation and their life cycles abruptly end. On a remote island, the only means of survival is to seize and consume any food that might be available in such a desolate place. Even if these lonely creatures survive to adulthood, finding a suitable mate is another daunting hurdle to overcome. For Hawaiian orchids rooted in a landscape largely devoid of insects, their best option is to self-pollinate.

Hawaii's rarest orchid

The rarest of the three orchid species native to the archipelago is the Hawaiian bog orchid, *Peristylus holochila* (syn. *Platanthera holochila*), with a mere three dozen individuals found in small clumps scattered across three islands. Most of the remaining individuals persist on the islands of Molokai and Maui, and only one plant - which is more than 40 years old - clings to survival on Kauai and almost perished in 1992 after Hurricane Iniki battered the site. Its reappearance from seasonal dormancy each year is a much-anticipated event on Kauai. When new leaves are first spotted emerging through the leaf litter, conservationists breathe a sigh of relief. The population on Oahu was not as lucky: *P. holochila* has not been seen on that island since 1938. Attempts to propagate the species in the 1970s and 1980s failed miserably, and in 1996, it was listed as a federally endangered species in the US. Only recently has it been propagated from seed by a team of dedicated college students in Illinois, but the process was slow and tedious. Given that so few plants remain, clonal propagation has been mentioned as one last option for preserving what remains.

How did this orchid arrive in Hawaii and from where, and why has it resisted efforts to propagate it? The answer to the first question may be rooted in the cold bogs of southern Alaska and the Aleutian Islands. According to Carlyle Luer, *Platanthera hyperborea* var. *viridiflora* is the prime suspect as the potential ancestor of

Steve Perlman (National Tropical Botanical Garden) walking on wooden planks in Alakai Swamp, Kauai, Hawaii (28 April 2011), where one *Peristylus holochila* individual orchid remains on the entire island.
Photo: Jon Letman/ National Tropical Botanical Garden

Peristylus holochila. They are virtually indistinguishable, producing up to 60 inconspicuous greenish flowers, tightly packed along an inflorescence that may reach 50 cm (20 in) in height. Luer hypothesised that the seeds of the Alaskan orchid may have arrived in mud caked on the feet of a migratory bird known for flying long distances over the Pacific Ocean. 'Seeds should not find transportation to the islands wanting,' he states, 'since the Pacific golden plover migrates annually between the cold bogs of Alaska and these high cool bogs of Hawaii.' Even if true, however, getting to Hawaii would only be a baby step towards establishment.

Regardless of how *Peristylus holochila*'s ancestors arrived in Hawaii, it didn't take long for speciation to occur, considering the islands' young geologic age. And now that it is there, the orchid finds itself clinging to survival by self-pollinating in isolation, alongside a motley assemblage of other endemic plants and animals that are just now learning how to co-evolve with one another. Given a few million years, Hawaii's bog orchid may team up with a hungry insect seeking a nectar reward. But surviving year to year is a struggle for most life forms that reach the archipelago. Only 18 specimens of the 'fabulous green sphinx moth' (*Tinostoma smaragditis*) of Kauai, for instance, have ever been collected by entomologists in the century since its discovery. It is now believed to be extinct, leaving *P. holochila* with one less option. With self-pollinating flowers and only a few dozen individuals to show for it, the Hawaiian bog orchid may exemplify a species teetering on the edge of extinction because the conditions needed for its long-term survival are no longer available.

However, extinction – like speciation – is a natural process. It has often been said that 99% of all species that ever lived are now extinct. Thus, it cannot be ruled out that *P. holochila* may have been on a natural path to extinction for quite some time, but adverse human-related factors may be accelerating its decline. This scenario is not unique to Hawaii's rarest orchid, as 9% of all native plant species in the archipelago have been declared extinct and over half (53%) remain vulnerable. Nellie Sugii at the University of Hawaii's Lyon Arboretum went so far as to suggest that Hawaii is, 'the endangered species capital of the world'. What we see happening in Hawaii is a microcosm of what lies ahead for orchid conservation on continents, namely that as intact forests are reduced in size, they become small archipelagos of their own.

Indonesia
'Situated upon the equator and bathed by the tepid water of the great tropical oceans, this region enjoys a climate more uniformly hot and moist than almost any other part of the globe and teems with natural productions which are elsewhere unknown. The richest of fruits and the most precious of spices are here indigenous. It produces the giant flowers of the *Rafflesia*, the great

green-winged *Ornithoptera* (princes among the butterfly tribes), the man-like orang-utan, and the gorgeous birds of paradise.'

Thus wrote Alfred Russel Wallace in his account of his eight years in the Malay Archipelago, which is made up of over 17,000 volcanic islands, most of them too small to appear in atlases, but including the large islands of Java, Sumatra, Sulawesi, Kalimantan in Borneo, and the western half of New Guinea. Despite encompassing just 1.3% of the world's land area, these islands are collectively home to about 10% of the world's flowering plant species. Little wonder that they have attracted botanists and orchid hunters in search of new treasures since the 19th century.

Deforestation has been, and remains, a serious problem, and the Indonesian Institute of Sciences has awarded funds to botanic gardens in Bogor, Cibodas, Purwodadi and Bali to set up a programme of exploration of the Indonesian archipelago, with the aim of making fresh collections of known species, and describing new species. Since 2021, all research centres in Indonesia have been merged into the National Research and Innovation Agency, whose goal is to establish a network of 32 botanical gardens. Botanists at Bogor Botanic Gardens have been exploring the archipelago since Dutch colonial times, and the Indonesian government has continued to support this activity since independence. Like her predecessors, Dwi Murti Puspitaningtyas' role is to explore natural habitats, thereby enriching the plant collections at the gardens. Her taxonomist colleagues are tasked with finding new species. We asked her how her team set about looking for new plants in such a large and diverse area made up of thousands of islands.

'Clear priorities must be established at the outset, and there must be a plan. The focus is on places and plants. In Java, wildlife is only found in forest conservation areas such as national parks, nature reserves and wildlife sanctuaries. The team chose to explore the primary and secondary forests that are still in good condition, especially those with a high density of trees that are rich in other plant species. If the goal is to target a certain species, the first step is to consult the preserved specimens in herbaria to determine where the plants were collected, and then visit that location. As an example, when looking for the endemic orchid *Paraphalaenopsis serpentilingua*, which is only found in restricted areas in Borneo, they were able to obtain information from local people or orchid hobbyists about where they were most likely to find those plants.'

A team usually consists of between six and ten people, including some researchers who are specialists (taxonomists) who study different plant families. Sometimes they are joined by ethnobotanists, plant identification experts with general plant knowledge, technicians who can handle plant collections, forestry officers in charge of protecting the forest, and local people who can act as forest guides. When looking for orchids, however, there is normally one researcher and one orchid expert.

Exploring the jungles of Bali, looking for orchids. Dwi Murti Puspitaningtyas is in the centre of the group.
Photo: Gusti Made Sudirga

Collecting and studying epiphytes presents challenges. Although they can sometimes be collected from fallen trees, on other occasions it is necessary for someone to climb a tree. Sometimes the technicians can climb the trees without help, or with local people assisting them. Fieldwork in the tropics can be physically demanding, sometimes involving trekking into the forest, crossing streams, and travelling by canoe. A young, inexperienced team must learn how to collect plants in a forest with many obstacles. Sometimes collectors must erect tents, be able to live in the jungle, and be mindful of hazards such as venomous snakes whenever they encounter them. They wear boots to avoid being bitten by snakes but, if bitten, they must know how to tie a tourniquet above the bitten part so that the venom is largely contained. The victim must then be taken to hospital immediately for treatment. Steep locations are generally avoided, unless members of the team have appropriate climbing skills. Local guides usually help to find a safe way to scale the karst limestone hills looking for *Paphiopedilum* species. Members of the team help each other so that they can climb cliffs, holding onto sturdy trees, before finally reaching the summit.

Before the young researchers are involved in a jungle exploration project, they are given training in the natural forest of Cibodas Botanic Garden (near Bogor Botanic Garden). They are taught how to collect plants properly, enter information into a database of plant collections, and make herbarium specimens of unidentified plant collections or new species, and they learn how to handle plants to keep them alive when they are sent from the forest to Bogor. Leaves are reduced, roots trimmed and old pseudobulbs are removed from epiphytic orchids. They are then placed somewhere in the shade where it is damp. The plants are misted by gentle spraying, taking care that they don't remain too wet, otherwise they will rot. The roots of terrestrial orchids are wrapped in wet tissue paper or damp moss (excess water is squeezed out), whereas small tender species, such as the jewel orchids (*Macodes petola* and *Anoectochilus* species), are placed in plastic bags with a pocket of air (first blow into the bag!) to maintain a humid environment.

Papua New Guinea
'The orchid flora of Papua New Guinea is one of the richest in the world, surpassed only by that of Colombia and Ecuador.'
Roland Schettler, 2021

The huge island of Papua New Guinea undoubtedly has the richest orchid flora within the Old World Tropics, with perhaps as many as 2,700 species, of which around 80 to 90% are endemic. In contrast to the other large islands in the Malesian region, much of the forest remains. Unfortunately, this is set to change. In August 2016, Roland Schettler and Phil Spence made a field trip to the highlands of Papua New Guinea to assess the distribution of several of the *Dendrobium* species in the section Oxyglossum. They were especially keen to find four species: *D. cuthbertsonii, D. masarangense, D. sulphureum*, and *D. vexillarius*. Spence had visited the area for more than 40 years and was astonished at how rare these plants had become. Once present in high numbers, the orchids are now scarce and difficult to find. Schettler and Spence reported that there appeared to be three main reasons for their demise: the occurrence of extremely dry seasons with no rain, and severe night frosts in 1997–1998 and 2015; the deliberate starting of grassland fires at Tari Gap, either wantonly or to hunt for food; and a small insect, possibly a scale or mealybug introduced accidentally by the Thai orchid industry, that is infesting orchid seed capsules and having a deleterious effect on populations. They concluded that members of the section Oxyglossum were extremely endangered, and likely to be lost from areas such as the Tari Gap within 10 years.

The Philippines
Comprising more than 7,500 islands spanning a total of 300,000 km² (116,000 mi²), the Philippines is one of the 20 most biodiverse archipelagos in the world. Its close proximity to orchid-rich South-East Asia and Indonesia explains why over 1,000 orchid species have been documented there, most of which are endemic. Today, as in the past, many new species await description, but there simply are not enough specialists available to clear the backlog. So the work of Jim Cootes and his research team at the Philippine National Herbarium is a daunting and never-ending task.

Perhaps the best-known orchid of the Philippines is *Vanda sanderiana*. It is known only from a couple of areas of the large island of Mindanao, and the story of its discovery by Carl Roebelin as told in *Frederick Sander: the Orchid King* reads like a detective story, in which Roebelin finds himself sleeping in a tree-house, only

A fine cultivated specimen of *Vanda sanderiana*.
Photo: Bob Fuchs

to be woken by a huge earthquake during the night. We are told: 'The tree-house was almost completely wrecked; the walls were in shreds and great gaps appeared in the walls and ceiling. Through one of these he saw a great spray of orchids such as he had never laid eyes on before.' Today, the main problem for this wonderful species is that it grows at mid-range elevations on huge hardwood trees that are sought-after by loggers. As loggers value lumber more than they value the epiphytic plants that grow on the trees they cut down, many are just left to die in the sun or to be eaten by whatever creature might take a fancy to them. As a result, in all probability, any remaining *V. sanderiana* only exist in remote places that have yet to be logged. These areas are usually under the control of anti-government separatists and are not safe places for anyone, particularly foreigners, to visit. Sadly, Jim reports that he has never seen this species in the wild, nor has he seen pictures of the orchid in its natural habitat. Fortunately, being the magnificent species that it is, there are countless seed-grown plants available for sale in the Philippines. Nurserymen from both Thailand and Hawaii have also worked tirelessly to preserve it.

Cootes works with a conservation group that is producing seed from the more popular species that can grow comfortably in the lowlands. Unfortunately, the beautiful miniature species that come from higher elevations will not grow in the heat of the lowlands and, because they are of purely botanical interest, no one seems interested in growing them from seed. This bias is not unique to orchids in the Philippines. If they cannot make a small profit from their seed-grown plants, their production will soon stop.

Palau (Micronesia)
Some of the same orchid species found in New Guinea and the Philippines made their way to the archipelago of Palau, located less than 1,000 km (620 mi) across the Pacific to the north and east respectively. This region of the globe is of interest to evolutionary biologists because it sits on the easternmost terminal point of Wallace's Line. This imaginary line was conceived in 1859 by Alfred Russel Wallace to mark a transitional zone between Asiatic plant and animal species to the west, and those found in Australia to the east. Organisms found on either side of the line clearly differ from one another, especially the land animals and birds. To the west – tigers and orang-utans. To the east – marsupials such as tree kangaroos. But for plants and especially orchids, artificial boundaries are no match for seed dispersal, and many species of orchids from both sides of Wallace's line made their way to the Palau archipelago. Palau's unique position raises the question of whether its flora and fauna are primarily Asiatic or Australian in origin. Considering that Palau was formed about 30 million years ago from volcanism, there has been ample time for different life forms from both sides of the line to colonise and evolve on many

of its islands. Among orchids, the number currently stands at 100 species, mostly epiphytes scattered on many small islands.

Today, these island 'states' are each ruled by different governors and traditional leaders. While most of these leaders grant access to orchid sites by researchers, others are more aloof, adding another layer of complexity to orchid conservation. In 2017, the Palau Orchid Conservation Initiative was formed to conserve Palau's orchids, in collaboration with the Smithsonian Environmental Research Center (SERC) and other partners. A SERC team was assembled, led by Benjamin Crain, and he soon found it necessary to meet each tribal leader in person, often multiple times, to gain access to a specific orchid site under their jurisdiction. This required considerable patience and tact. Most were cooperative, but many had temporary appointments, meaning that permission granted one year could be denied the next.

VIEWPOINT

Orchids in Madagascar
by Lawrence Zettler

'Your lovely island of Madagascar is rich in plant and animal biodiversity, yet this treasure is especially threatened by excessive deforestation, from which some profit. The deterioration of that biodiversity compromises the future of the country and of the Earth, our common home.' Pope Francis, 2019

Madagascar is the world's fourth largest island. It was part of the Indian subcontinent until about 90 million years ago, when India and Madagascar separated. India drifted north and collided with Asia, forming the world's highest mountain chain, the Himalayas, while Madagascar was left isolated 400 km (250 mi) off the south-east African coast, separated by the Mozambique Channel. What we see today is a place brimming with endemic species, exemplified by lemurs and 90% of its native orchids. Although geologically of Indian origin, Madagascar's life forms have closer biological ties to those in Africa.

Madagascar has more orchid species than the whole of Africa combined – around 1,030 at the last count – with the number steadily increasing as orchid biologists continue to explore the rugged landscape. Unfortunately, the island's orchids are threatened by slash-and-burn farming (locally known as *tavy*), wood harvesting (for firewood), and mining. Specialists throughout the world are racing to study and save as many species as possible, as quickly as possible.

In 1986, Phillip Cribb spearheaded the effort to study and protect Madagascar's unique biodiversity by establishing Kew's only permanent overseas research station, the Kew Madagascar Conservation Centre (KMCC), based in the capital, Antananarivo. Today, KMCC employs 40 Malagasy scientists, including Landy Rajaovelona, one of several specialists familiar with the island's orchids. In 2012, with the help of KMCC and its staff, a research team led by Kew's Viswambharan Sarasan and funded by the Sainsbury Orchid Conservation Project was organised to study and conserve the orchids of Madagascar's Central Highlands over a five-year period. The story that follows describes our personal experiences in Madagascar spanning several separate trips.

SAVING ORCHIDS

(opposite)
Angraecum longicalcar – notice the extremely long nectar spurs.
Photo: Andy Stice

(above)
Andy Stice (Laboratory manager, Illinois College) seed baiting *Angraecum longicalcar* **in Madagascar with seed prior from Kew.**
Photo: Lawrence Zettler

(right)
Lawrence Zettler peparing slides for seed baiting.
Photo: Andy Stice

(above left)
Slide mounts containing seeds of *Angraecum longicalcar* before being buried adjacent to a mature individual of the same species rooted in a rock crevice in Madagascar, photographed in April 2013.
Photo: Andy Stice

(left)
A mural hand-painted by local residents where *Angraecum longicalcar* grows naturally as a lithophyte in Madagascar. Photo taken in April 2013.
Photo: Andy Stice

All (dusty) roads lead to the lab

In April 2013, Sarasan's research team travelled to Madagascar at the tail-end of the rainy season. The team consisted of Jonathan Kendon and Kazutomo Yokoya, from Kew, and me, Lawrence Zettler, and my colleague at Illinois College, Andy Stice. Landy Rajaovelona and Gaëtan Ratovonirina, specialists from KMCC, completed the team. They arrived hoping to obtain fungi from orchid roots, especially those from young seedlings.

After a day's drive over dirt roads with potholes deep enough to swallow a small car, the dust-covered SUVs pulled into a charming family-owned motel located near Analabeby, a small village where children were playing as the sun set. A young teenage boy walked along the street, proudly carrying an AK-47 machine gun, its ammunition cartridge long since spent. On the outskirts of town, growing just a few miles away in an area stripped by mining, was *Angraecum longicalcar*, one of Madagascar's rarest and best-known endemic orchids. This spectacular orchid once grew here in large numbers on the hot and dry rocky hills of the Itremo Plateau east of Ambatofinandrahana. Today, the species is reduced to a handful of plants, all restricted to a small area in full sun on four separate marble outcrops or inselbergs. Finding seedlings of *A. longicalcar*, and the fungi that they require in nature, was of utmost importance.

Early the next morning, we left the motel and drove down a track to the *A. longicalcar* site. Before the ignition could be turned off, young bare-footed children arrived at the scene eager to meet the team, who probably seemed like aliens from another world. As we unpacked supplies and walked over the rocky terrain in search of orchid seedlings, the children followed in our footsteps. After several hours of scouring the landscape searching for seedlings between and on rock surfaces, none was found.

The site had previously been visited by David Roberts (assisted by Phil Seaton) in October 2005, when Roberts had sampled leaf material. It had been difficult to be certain just how many individuals were present as the plants were growing in large clumps, with each shoot connected by a mass of tangled rhizomes, and the aim had been to determine how many distinct clones were present. Samples were placed in individual resealable plastic bags containing silica gel as a drying agent, and their GPS co-ordinates recorded. Some plants still retained battered labels – remnants of a previous investigation by Rajaovelona, who had been investigating the species' population biology and reproductive success. She had discovered that the species is rarely pollinated, so fruit production was low. The plant appears to be strictly resource limited, and only flowers when it builds up sufficient energy. Indeed, Roberts was unable to find any seedlings, just one dead flower spike, and one seed capsule.

At the time of the visit, this orchid seemed on the edge of extinction in the wild. Recently one plant had been illegally harvested by a collector. The thin tufty grass of the region was regularly burned by the local herdsmen to generate fresh new shoots for their zebu cattle, and a recent fire had gotten out of control, spreading through part of the orchid population. In addition, the far end of one of the outcrops was being mined for its marble.

Since that time an environmental association called 'Mihary Soa' had been created by the local population. High above the site, on a rocky exposed cliff, is a hand-painted fresco that depicts the orchid that the local community now cherishes. The display is heartwarming, not just because it expresses concern for the well-being of an orchid, but because everyone, young and old alike, recognizes the significance of this coveted species. Essentially, the entire community serve as agents of conservation under a 24-hour watch. When Sarasan's team arrived at the *Angraecum longicalcar* site, the same hand-painted fresco, painted over a decade ago, greeted the researchers as bright and bold as if it had been painted yesterday.

Cyclone-driven discoveries
A final trip to Madagascar took place in January 2015, in the middle of the rainy season. Apart from the landscape being a little greener and the roads less dusty, everything looked much the same as the team departed Antananarivo. After travelling south for a good part of the first day, we decided to stay overnight in a village that had electricity, food, and gasoline needed for the days ahead. On the village outskirts was a small unpaved road that served as one of the region's only west-to-east arteries, used by heavy trucks carrying goods and lumber back and forth. At several points along the road, bridges had been erected from scraps of wood across streams and small rivers choked by the silt of erosion resulting from deforestation. Both the two SUVs and their occupants would be tested to their limits.

During dinner at the hotel, Jacky Andriantiana from Tsimbazaza Botanical and Zoological Park arrived looking worried. A tropical cyclone had formed in the Mozambique Channel and was headed directly for the study site. There was a storm-proof shelter available, assuming the team arrived before nightfall the following day, when the cyclone was expected to hit. If the team didn't reach its destination before dark, however, there was a risk of being robbed at gunpoint by bandits. Jacky's facial expression was not reassuring.

Early the next morning, both SUVs headed out as the rain began to fall. After leaving the village, the road became an angry, slithery serpent of boulders embedded in clay muck that, in places, disappeared into potentially lethal potholes. As the wind and rain intensified, the SUVs lost traction and came to a standstill four times during the journey. The fourth time occurred at sunset just shy of a bridge over a river. Abandoning the trip and driving back

Digging out an SUV in the Central Highlands of Madagascar for the 4th time at the onset of a tropical cyclone during the rainy season in 2015. Tents were erected shortly after the photograph was taken.
Photo: Lawrence Zettler

was not an option, so the team pitched tents and hunkered down for the night.

Around midnight, the storm passed directly overhead with such force that the corners of the tents flapped up and down like some large waterfowl trying to fly. With each passing hour, the winds began to gradually subside. At daybreak, we discovered that the bridge that provided access to orchid-rich habitats beyond had been washed away, with only a few wooden pillars projecting above the water line remaining. With both SUVs stuck deep in mud, we walked to the nearest town 7 km (4.5 mi) away, while Andriantiana remained behind to coordinate the extraction of the vehicles.

By afternoon, the team had arrived in a small village, soaked and muddied. A small hotel served as a place to dry off for the evening. On my unmade bed were water-soaked CITES permits that were wilted and stained blue with ink. Above the bed was a mosquito net tied into a knot ready to be deployed to keep mosquitoes, and the malaria they carried, at bay. In the hotel lobby, Andriantiana was back from digging out the SUVs, and ready to discuss Plan B. The group decided to collect orchids from sites where bridges still functioned, targeting several lovely *Cynorkis* species in full bloom, and lithophytes on rocky outcrops, such as *Aerangis ellisii*.

During a lunch break, sitting on a large volcanic boulder overlooking a picturesque valley, Andriantiana and I reflected on what had occurred during and after the cyclone. Andriantiana asked: 'You are from America?' I nodded. 'The last American that came here with me was a well-known scientist. When he returned to the States, he e-mailed me and said he was sick, then he stopped communicating with me. I found out later that he had died from bilharzia, which he had contracted after walking through rice paddies here.'

This sad tale illustrates just how hazardous field trips can be for botanists. Our trip also epitomised the situation that wildlife faces in Madagascar. Located down a lonely highway surrounded by cassava fields was a small remnant forest which was home to lemurs and rare terrestrial orchids. At the forest's edge was a small house powered by solar energy. A young woman carrying a small child greeted the team and walked barefoot into the forest, pointing out its orchids. Almost every tree had at least one scar inflicted by an axe or a machete. Back at the house, after the team collected their samples, the woman was paid for the tour and given a sizeable tip. Behind her was a large pile of cassava tubers heaped on the porch. The only reason why the forest had not been cut down and replaced by cassava was because the forest earned more money per hectare from ecotourism than the surrounding agricultural fields.

The mysterious lithophytes

Although *Angraecum longicalcar* and many other Madagascan lithophytes have been successfully cultivated from seed in the laboratory for many years, their natural germination requirements remained an enigma. Dry, exposed rock hardly seemed a good place for mycorrhizal fungi to flourish. The question was: where do these orchids germinate? Sarasan's team made an important discovery that shed some light on the mystery.

Following the cyclone, the team climbed a remote rocky outcrop near the village of Mahavanona. Clinging to the numerous marble rocks were a handful of *Aerangis ellisii*. A number consisted of dead root masses that were slowly decomposing, their tops and apical meristems decapitated by an apparent act of vandalism. Kendon ventured onto one large rock crevice filled with decomposing leaves. Carefully dipping his bare hand into moistened bits of crumbled leaves, he pulled out six small, white ovoid spheres less than one centimetre in length that resembled orchid protocorms. Upon closer inspection of the crevice, a long brittle root could be seen adjacent to where the protocorms were found.

Laboratory results confirmed that the spherical objects were indeed tiny leafless orchid seedlings (protocorms), and that they originated from *Aerangis ellisii*. The protocorms and the root also harboured a fungus that demonstrated a remarkable ability to germinate seeds in the lab, confirming that it was a mycorrhizal associate. As a result, Sarasan's team hypothesised that lithophytic orchids begin their lives in rock crevices, perhaps during the rainy season when organic matter begins to decompose. Over time, some of the roots that form remain anchored in the crevices, while others branch out and colonise the bare rock surfaces. The discovery of *A. ellisii* protocorms was a fitting way to conclude the Madagascan experience. The trip served as a reminder that scientific discoveries in the modern age still await those who are willing to venture abroad, despite all the potential risks and setbacks.

Australia: orchids in peril down under

Aside from Madagascar, nowhere harbours as many unique and wonderful plants and animals as Australia. Like Madagascar, Australia is remote, surrounded by water, and its creatures have lived in isolation for millions of years. Many of its native mammals are exceptional. The echidna and the platypus lay eggs, for example. Marsupials, such as the kangaroo, koala, and bilby, give birth to tiny young that wiggle up and into their mother's pouch. Are the orchids in the 'Land Down Under' equally unique? Absolutely.

Although largely arid, Australia is home to more than 1,700 orchid species. Around one quarter are epiphytes, restricted to a narrow region along the country's eastern margin in the moist tropical forests of north-east Queensland. On the opposite, drier end of the continent, in the state of Western Australia, terrestrial orchids reign supreme, with more than 300 species. As such, this region has been designated as an orchid 'hot spot' of global significance. It is here that some of the most unique orchids are found, assuming one is lucky enough to find them. The diminutive hammer orchids (*Drakaea* species) are known for pollinating their flowers by deceiving male wasps seeking sexual pleasure (pseudocopulation). The 'underground orchid', *Rhizanthella gardneri*, exemplifies extreme specialisation, evidenced by its exclusively subterranean existence and lack of either green leaves or any obvious roots. Ants and termites have been implicated in pollinating its flowers, which remain underground or gently break the surface. The orchid also relies entirely on a fungus connected to the roots of woody shrubs (*Melaleuca* species) to supply its food, and its seeds will only germinate when re-connected to the shrubs' roots via the fungus. According to Kingsley Dixon, who was the founding director of science at Kings Park and Botanic Gardens in Perth, the population of underground orchids in Western Australia numbers fewer than 100 individuals, which are threatened by land clearance for agriculture. 'Kings Park has secured seed and mycorrhiza for long-term storage', he says. 'However, much more needs to be done to halt habitat degradation and to fund reintroduction programmes.'

The seeds of underground orchids have been described as being 'like ball bearings', and mammals, such as wallabies and bandicoots are the prime suspects for their dispersal. In eastern Australia, highway construction threatened a population of another endangered *Rhizanthella* species (*R. slateri*). It was only saved from the bulldozers when the highway was diverted. Some of the plants were also successfully relocated. It seems that no place on Earth is free of 'progress' and no orchid is safe, not even those that escape reality by disappearing beneath the ground.

VIEWPOINT

**Life after death for orchids
in a biodiversity hotspot**
by Kingsley Dixon

My life in orchids has been enriching, endlessly rewarding and with many twists and turns. Solving the propagation of recalcitrant species, coming up with new and faster methods, and sharing experiences with the orchid world of researchers and enthusiasts means that there is never a dull moment. But with the extinction clock ticking ever more loudly, at times the challenge is so daunting it is difficult to know if we can make a difference. But that same enthusiasm that drove me as a young child to cherish our orchid heritage still drives me today. We can and will turn orchid conservation around. We owe it to those generations to come to share the joy of the splendid beauty of orchids in the wild.

I can't recall exactly when, but somewhere in the wilds that surrounded our property in south-west Australia, orchids crept into my life. Growing up in the 1960s, among the wild plants, I remember my granny telling about rattle beaks, blue lady orchids, spider orchids, and cowslip orchids. The names alone started a budding botanical urge that awakened dormant interests embedded in my DNA. Soon I was building shade houses and, though still at school, I started reading, collecting, and growing bush orchids. My classmates were puzzled, but somehow sympathetic to what must have appeared an almost religious zeal for native orchids. So was born my lifelong love of orchids, their botany, culture, and conservation.

It wasn't always plain sailing. There was no Google or internet where I could find like-minded orchidophiles, so I pushed on with reading the scant literature and imagining names for many of the orchids I would see on my endless rambles. My parents watched on, and very soon they were hooked too. Not long after, cymbidiums, paphiopedilums, and native dendrobiums started to arrive at home, and more shade houses and a greenhouse were constructed, with weekends spent visiting other orchid growers. Journals arrived from local orchid societies advertising orchid meetings and shows: magical, mystical places with show benches heaving with wondrous orchids I had no idea existed.

That was then. Bushland was seemingly endless, as were the rich and ever-present bushland orchids. Down the road or across the street we could always find orchids. University beckoned and my bushland rambles became less frequent and more difficult as the relentless expansion of urban Perth swallowed up all my childhood wild haunts. Today, a freeway exists right where I found my first magnificent white spider orchid (*Caladenia longicauda*), and my favourite patch of the early blooming banded greenhoods (*Pterostylis sanguinea*) is lost under a sea of houses.

A turning point in my love of orchids came when I was introduced face-to-face with the type specimen of the Western Australian underground orchid (*Rhizanthella gardneri*). I had bugged my aunts with contacts at the State Herbarium to see the orchid. Till then, all I had were the shadowy sketches of the orchid in my botanical bible, Rica Erickson's 1951 publication, *Orchids of the West.* Peering at the orchid's capitulum (inflorescence) suspended by cotton threads and preserved in alcohol, I was transported to 1929 when the species was discovered during farmland clearing. What I would find out later in life is that every subsequent discovery was also an extinction, as plants were only found during agricultural clearing. At that time, I resolved that somehow, I would become a botanist and work on orchids. How, when, and if I could make a living seemed irrelevant, and so off to university I went.

My school careers advisers thought my interest in plants meant I wanted to grow wheat, so I enrolled in the University of Western Australia Faculty of Agriculture. But then I saw a building with Botany emblazoned on the front. Enrolments were hastily changed and by my second year I was displaying my prized collection of potted native orchids in the the lobby of the Department of Botany. I had finally found like-minds. Bushland rambles with university researchers took our classes to amazing places with even more amazing orchids. My horizons expanded and life was good.

When I was accepted to undertake a PhD I was asked by my professor what I would like to research. I replied 'Orchids'. The response was shattering and clear. 'You can't grow native orchids, so you can't do a PhD on them'. Here I was speaking to Professor John Pate, one of the world's leading legume physiologists who just two years earlier had arrived in Perth from Queens University in Belfast. What was I expecting as a response! My PhD was not to be on my beloved orchids, but on native plants with underground storage organs. But orchids have underground organs, so I had an out.

With the PhD firmly under way, in my spare time I started to assemble literature on terrestrial orchid propagation. I wrote to the doyen of orchid mycorrhiza in Australia, Dr Jack Warcup, as he had published a series of papers on symbiotic germination. I received a response I could hardly believe. Three vials of the symbiotic fungus for *Pterostylis vittata* (syn. *P. sanguinea*). I started to germinate native orchids for the first time, but only this species. Isolating orchid mycorrhiza was enshrined in Warcup's complex methodology, but it eluded my early attempts.

But then, an advertisement for a botanist appeared in the department. Kings Park and Botanic Garden, Perth, and the World Wildlife Fund (now the World Wide Fund for Nature), were seeking a qualified person to search for the Western Australian underground orchid that had been presumed extinct for 20 years. I was completing my PhD so could work part time. I could hardly believe my luck when I was called for an interview. But in my unbridled enthusiasm for the two-year position I was totally unprepared for the first, and most obvious question. 'How would you look for a plant that never appears above ground level?' I was stumped. Would I plough up the last small pieces of remnant bushland? Then it came to me. A lecture on mycology talked about how the French used rakes to unearth truffles. I responded, 'I would use truffle rakes'. The panel were clearly impressed and shortly after we were off searching for the orchid.

We found nearly 180 plants in that first year, but they were restricted to two tiny bushland reserves. When that post was about to finish, I joined Kings Park as a botanist in 1982 and

The underground orchid, *Rhizanthella johnstonii*.
Photo: Kingsley Dixon

established symbiotic methods based on Mark Clements and R. K. Ellyard's seminal work. My research life with orchids had finally arrived.

Within a few years, the orchid research team at Kings Park had grown, sponsorship and grant funding flowed and international collaborations blossomed with the Royal Botanic Gardens, Kew. The International Orchid Conservation Congress started as a means for orchid conservationists to have their own meeting (rather than being consigned to the periphery of larger plant conservation conferences). In order to address the grim task of reversing the threatening tide of orchid extinctions we needed to combine our global capacity, and the meeting continues today to be the premier venue for orchid conservation science and practice.

Orchid research at Kings Park grew at a phenomenal rate, with outstanding research students joining the team over the 32 years that I was Director of Plant Science. Always keen on science and making a difference to conservation through communities and schools, we developed our orchid house after fighting to save a series of former display glasshouses from being demolished. With nearly 100 species in the collection, we finally had a home for our programmes and our plants, and our orchid volunteer team worked tirelessly with the Friends of Kings Park to enrich the collection as a seed orchard for propagation. A highlight was an idea I had that we needed to reach out to more people about orchids through a user-friendly propagation kit. So was born the Orchid Kit, which had a packet of pasteurised mulch, and seeds of the locally grown donkey orchid (*Diuris magnifica*) and carousel spider orchid (*Caladenia arenicola*). They sold out within weeks and, 20 years on, plants are still being grown from those original kits.

Victoria, Eastern Australia
'Anyone who works with threatened species needs to be an optimist.'
Noushka Reiter, 2023

On the other side of the continent, Noushka Reiter manages the Orchid Conservation Program at the Royal Botanic Gardens, Victoria. Based firmly on scientific principles, the aim of her project is to achieve the long-term re-establishment of viable and self-sustaining populations of many of Australia's highly endangered orchids. To maximise the potential for successful translocation of species back into their natural habitats, Reiter's team applies a whole-ecosystem approach to conservation, studying not only the orchids' life cycles, but also their associated mycorrhizas and pollinators. The orchids are propagated using symbiotic techniques to ensure the inclusion of appropriate mycorrhizas when seedlings are planted in the field. Reiter has pioneered new methods to study food-foraging pollinators, using cultivated plants as 'bait'. This enables the orchids to be transplanted into the wild only where the presumed pollinators are present.

Many, perhaps most, conservation activities around the globe are fueled by 'the smell of an oily rag' depending largely on grants, donations, and the participation of volunteers. Citizen scientists, including retired scientists, assist in both the lab and the field, and help to promote wider public awareness of the plight of many of Australia's native orchids. Volunteers from the Australasian Native Orchid Society (Victorian Group) have been a key community partner, annually contributing more than 2,500 skilled hours of work. Training for the volunteers, who arrive with varying levels of skills and experience, provides background knowledge on the species they are working with. Learning continues during the activities, and sharing of knowledge between volunteers and stakeholders is ongoing. Activities at the laboratory and nursery of the Royal Botanic Gardens, Victoria include seed germination counts, data entry, flasking seedlings, and potting orchids. Field activities comprise population surveys, monitoring, planting for translocations, weeding at important sites, and surveys for orchid pollinators. Volunteers are taken back later to the translocation sites to assist with monitoring and post-translocation maintenance, allowing them to see the results of their hard work.

Reiter's research has led to an increase in both populations and total numbers of 14 species of endangered orchids. For instance, the spider orchid, *Caladenia audasii*, which is endemic to Victoria, was at one time reduced to five known individuals remaining in the wild. This species was cultivated to its flowering stage for the first time, and different individuals were hand-pollinated to obtain seed for propagation. On the evening of 23 August 2023, Reiter was awarded Australia's prestigious Eureka Prize for Botanical Excellence in recognition of her outstanding orchid conservation work and its outreach throughout the country.

Based on the propagation methods developed by Reiter, the project is underpinned by one of Australia's most significant *ex situ* collections of threatened plants. At the time of writing (April 2023), this number included more than 20,000 individuals, representing 212 species, including populations of 90 endangered orchid species. The collection preserves genetic diversity, provides an opportunity for seed banking, is a resource for research, and maintains a legacy for the long-term conservation of Australian orchids.

Marc Freestone (formerly at the Royal Botanic Gardens Victoria Orchid Conservation Program) tells us that native grasslands in south-east Australia host a diversity of endemic orchids, most of which are highly threatened due to widespread land use change in their native grassland ecosystems. Unlike woodland, heathland, forest, or alpine habitats, which largely don't require management (apart from herbivore and weed control), native grasslands require constant management. They occur mostly on roadsides, rail reserves, aerodromes, and in old cemeteries and golf courses, because almost all habitat on private land has been ploughed, fertilised or over-grazed. These remnants are largely managed by the state. Cemeteries are usually managed by a local cemetery trust, and most golf courses are private or managed by local government. It can be difficult to convince Australian land managers of the need

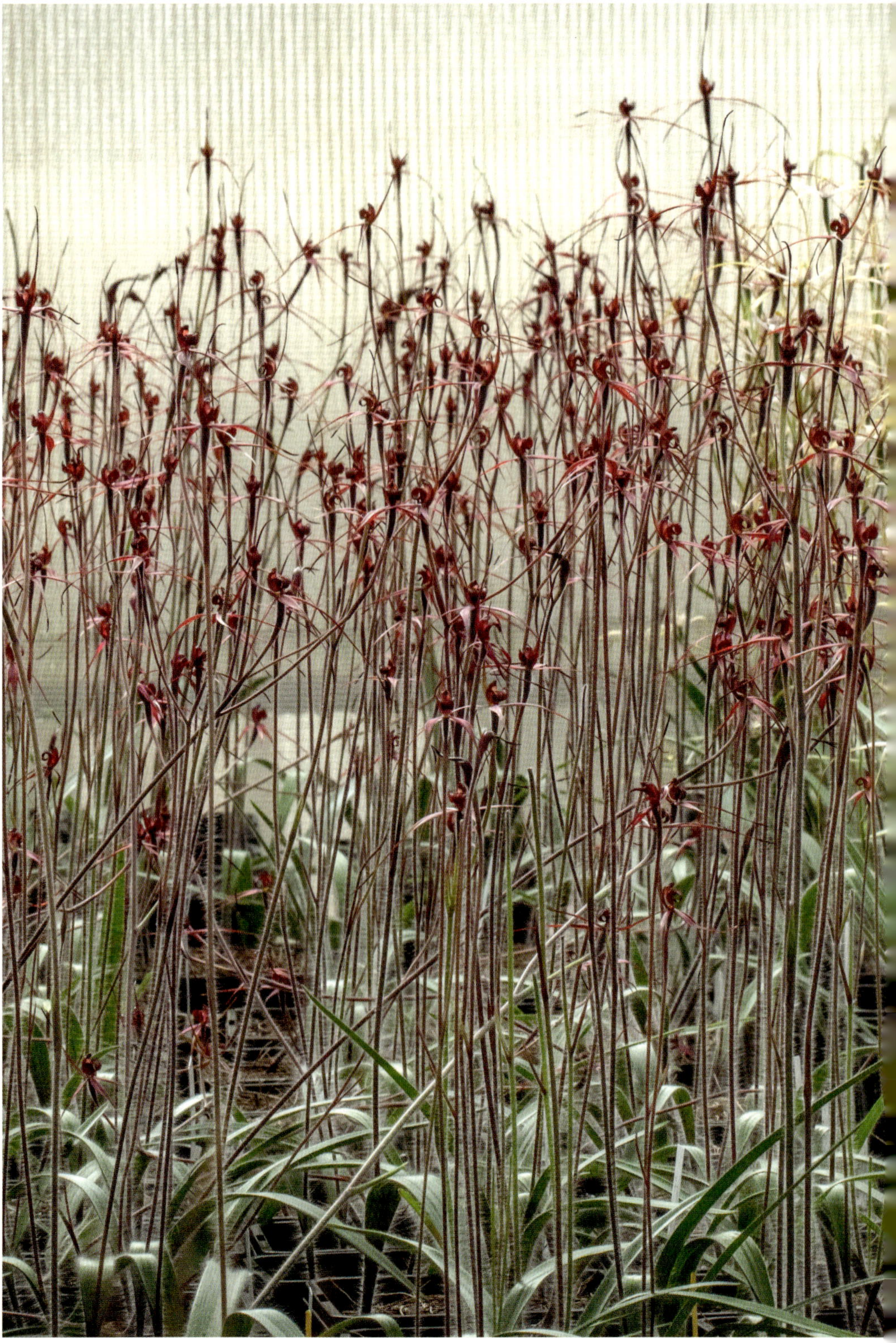

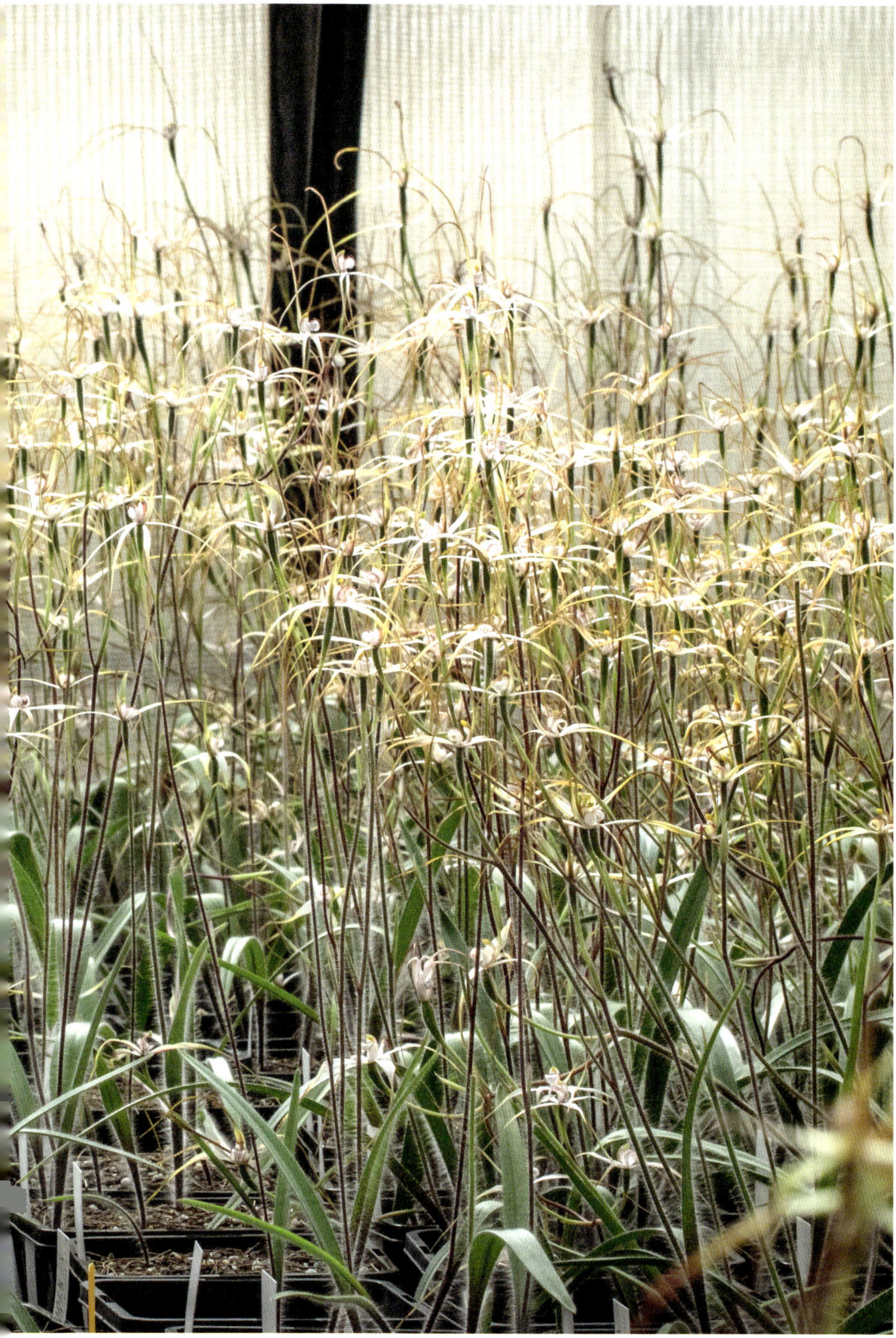

(previous pages)
Populations of *Caladenia cretacea* **and** *C. cruciformis* **grown from seed.**
Photo: Noushka Reiter

(left)
Volunteers planting orchids in the bush in the State of Victoria, Eastern Australia.
Photo: Noushka Reiter

for regular burning, instead, herbicides are used to keep down the vegetation, with all-too-predictable consequences for the orchids. Some roadsides are looked after by local fire brigades, but they are rarely interested in conservation, and chiefly burn them to reduce fire risk to adjacent farmland. By contrast, prescribed burning of tall-grass prairies – an activity practised by Native Americans for thousands of years in central North America – has recently been shown to benefit the fecundity of two threatened *Platanthera* species in the US.

The sunshine diuris
'The area where I was brought up is still being deforested even today. To see the bush where I used to play, in my case dry eucalypt woodland with a low understory of a few acacias, being stripped back for mining and agriculture is very sad.' Zöe Smith, 2009

In the past 150 years, more than 60% of Victoria's land has been cleared for urban development or agriculture. According to the IUCN 2001 Red List, 208 of the state's 450 orchid taxa were either threatened or extinct. *Calochilus richiae* has been reduced to fewer than 20 plants, and *Caladenia amoena*, *Pterostylis aenigma*, and *Paracaleana disjuncta* are known from single populations of fewer than 100 individuals.

At one time, the sunshine diuris (*Diuris fragrantissima*) was locally abundant in the kangaroo grass (*Themeda triandra*) grassland. Today, less than 1% of this habitat remains. More than

99.5% of Australia's native grasslands have been destroyed, with the remnants restricted to small, fragmented areas, such as roadside and rail reserves that continue to be threatened by disturbance and invasion by weeds. Strips of vegetation along railway lines were once home to many relict orchid populations – the occasional trackside fires caused by stray sparks from steam trains probably helped to maintain orchid habitats by removing scrubby vegetation.

Before beginning her study of *Diuris fragrantissima* at the Australian Research Centre for Urban Ecology of the University of Melbourne, Zöe Smith had to ensure that it was a 'good' species. Was it sufficiently different from the related *Diuris punctata* to merit species status, or was it just a variety? Was it worth the conservation effort? With limited resources, no one wants to spend time on an orchid that turns out not to be endangered after all. With such a small population, plants couldn't be removed from the wild. To confirm that *D. fragrantissima* was indeed a true species, Smith chose 80 floral and vegetative characters, and spent up to 12 hours a day in the field for about 30 days measuring randomly selected plants. When data like these are presented in a graphical form, individuals with the greatest similarity group together, while those that differ form separate groups. *D. fragrantissima* formed a group clearly related to, but distinct from *D. punctata* – a finding supported by genetic fingerprinting.

In the 1980s, a number of plants had been removed from the wild to establish an *ex situ* collection. Rather than confining *Diuris fragrantissima* to an 'orchid zoo', Smith wanted to establish a second population in the wild. All previous attempts at reintroduction had, however, failed. The first step involved germination using a mycorrhizal fungus, which she isolated from the wild population. It was also possible to germinate the seeds successfully without the aid of a fungus, using Kevin Western's patent medium. But as it turned out, all such seedlings rapidly became infected with a fungus in the nursery, and DNA sequencing revealed this fungus to be almost identical to that found in the wild. How the fungus spread between pots remains a mystery. It could have been in the potting compost, for example, or perhaps transferred by splash-over when the plants were being watered.

In 2004, more than 700 *Diuris fragrantissima* orchids were reintroduced into the wild. Those with large tubers that were planted during the spring or autumn were most likely to survive, probably because they could tap into the plentiful source of carbohydrates (starch) and water in this storage organ. Adding a fungus coupled with soil tilling also proved beneficial during the spring, when the clay soils were less prone to compaction by sheep. After the first two years, half of the plants with tubers flowered, which was very encouraging. However, survival and flowering mysteriously declined thereafter, which demonstrates a need to understand more about the species' biological requirements.

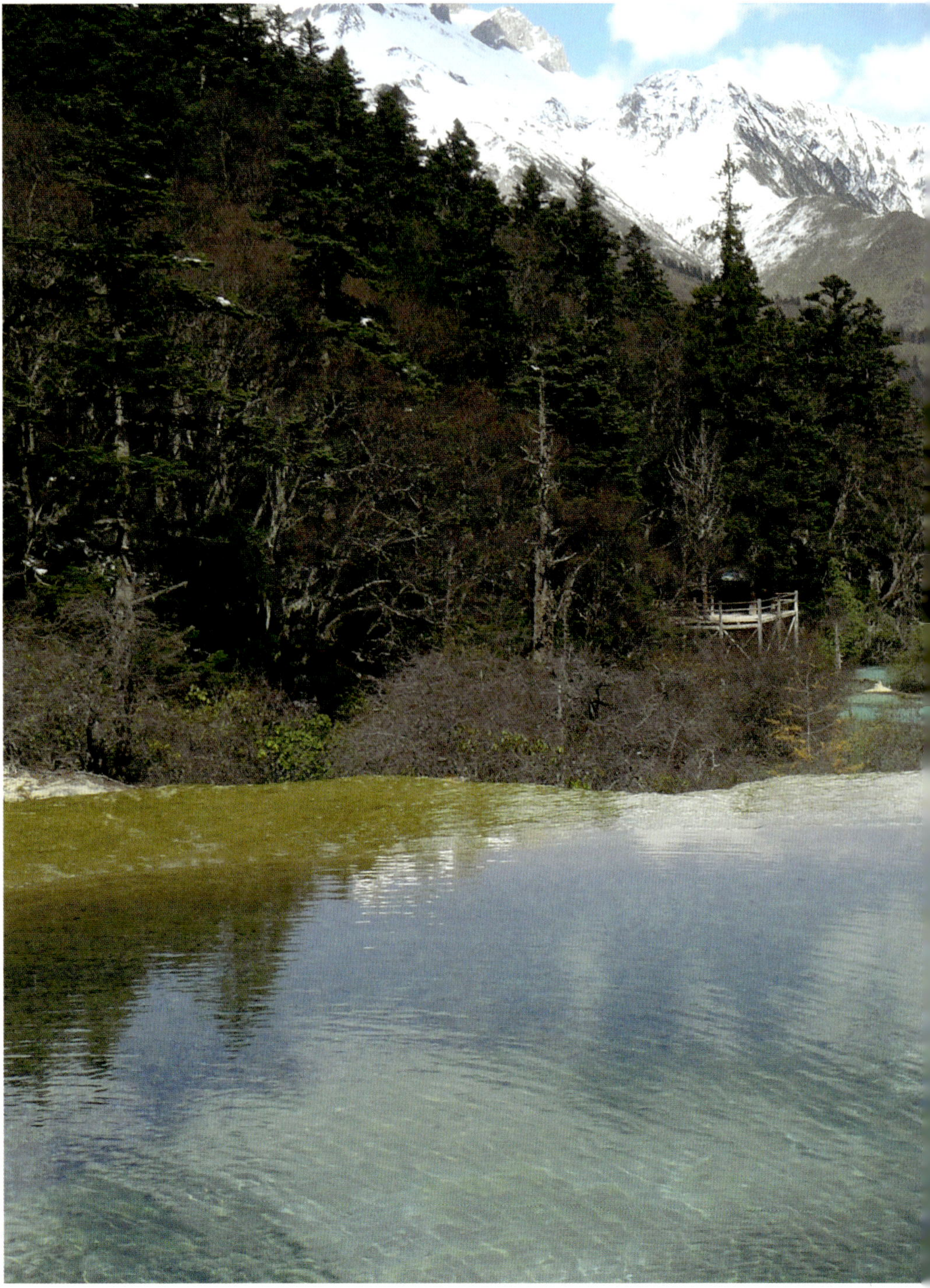

6 Where are we most likely to find new species?

After more than 200 years of exploration, the rate at which new orchid species and natural hybrids are being described has not declined, despite ongoing habitat destruction. Undoubtedly, many new species await discovery, and it is reasonable to assume that most (but not all) will be found in biodiversity hotspots within the tropics. To qualify as a biodiversity hotspot, a locality must have at least 1,500 endemic species of vascular plants, and have lost at least 70% of its original natural habitat. Based on these broad criteria, established by Conservation International, some 36 biodiversity hotspots have so far been identified, comprising just 2.5% of the Earth's land surface.

Between 1972 and 2002, at least 280 new species of orchid were described worldwide each year on average. In 2019, 288 new orchid species were scientifically named for the first time. There are probably hundreds of new orchid species waiting to be discovered in Colombia alone. In Ecuador, Lou Jost continues to find many new species in the cloud forests around Baños. Today, most discoveries involve species that have tiny flowers – so-called 'botanicals' – which are important in themselves, but of little horticultural interest. Others, however, are spectacular. In 2000, Franco Pupulin discovered *Dracula inexperata* while walking along a familiar trail in Tapantí National Park in Costa Rica. The scarlet slipper orchid, *Phragmipedium besseae*, from Peru and Ecuador, was first described in 1981, and the enormous purple blooms of *Phragmipedium kovachii*, which is native to the cloud forests of northern Peru, created a sensation when it first came to the attention of the orchid world in 2001. In 2020, *Gastrodia agnicellus* from Madagascar made the news when it was described by Johan Hermans at the Royal Botanic Gardens, Kew as possibly the world's ugliest orchid. It was subsequently included in Kew's 2020 top 10 species of plants and fungi new to science.

Many mountain habitats in Ecuador remain unexplored. The discovery of three new species of *Lepanthes* in Ecuador demonstrates the importance of continuing the botanical exploration of a mega-diverse and incompletely inventoried country. *Lepanthes microprosartima*, for example, was found in evergreen montane forest on the western slopes of the (periodically active) Pichincha volcano, where it is endemic to the Yanacocha and Verdecocha reserves. Only 40 individuals of a second *Lepanthes* species, *L. caranqui* (named in honour of the local Caranqui culture) were found in evergreen montane forest in Pichincha and on roadside embankments above the tree line in Imbabura. The third *Lepanthes*, *L. orolojaensis*, was found in just one locality on the border between Oro and Loja provinces. Its original habitat is constantly threatened by cattle ranching, removal of shrubs for firewood, and rapid fragmentation of the surrounding landscape due to fires and plantations of exotic pine trees (*Pinus radiata*).

(previous pages)
Northern China is a centre of biodiversity for slipper orchids (*Cypripedium* species), and a number of new species have been found in the area of Huanglong in recent times.
Photo: Philip Seaton

In Colombia over the past decade, Marta Kolanowska has discovered and described a population of 30 plants of a new species, *Telipogon diabolicus*, on the rim of the Sibundoy Valley. They grow epiphytically in wet, dwarf montane forest on the edge of the páramo (tundra). Immediately listed as being critically endangered, the species created a lot of media interest, probably because of it name. The petals are rose pink (telipogons are typically yellow), and the centre of the flower is dark rose, resembling the head of a little devil. All the media interest in the discovery is, however, somewhat double-edged. The publicity for orchids and their conservation represents a positive, but the location of the plants along a road in public view means that the population may be vulnerable to collectors.

One reason for the ongoing disagreement on how many species exist reflects, in part, the fact that there is a shortage of taxonomists in the tropics. For example, 1,264 orchid species have been identified to date in Bolivia, but the actual number may fall between 2,000 and 3,000 considering that so many species have not yet been identified. In neighbouring Ecuador, some new species have been discovered from illegally collected orchids seized in airports. Because these orchids are collected without any trace of their exact origin, they cannot be returned to the wild. Instead, they are taken to specialists at the University of Cuenca where they later flower in captivity, revealing their unique taxonomic identity. How many will turn out to be endemic to that country? Who knows? In common with many of us, ecologists like a cold beer and a hot shower at the end of their working day, and so it is likely that many orchids remain to be discovered in the more difficult-to-access, inhospitable places.

New populations of known species are also still being discovered on a regular basis. A formerly unknown population of the rare and endangered *Laelia dawsonii* f. *dawsonii*, for example, was recently found in Jalisco, Mexico. Similarly, a new population of *Mexipedium xerophyticum*, first discovered in Mexico in 1988, was found in 2009. On the other side of the globe, a new population of 250 donkey orchids (*Diuris callitrophila*) recently surfaced in New South Wales, Australia. The previous population had plummeted to 1,000 individuals. Meanwhile, Hong Jiang from the Yunnan Forestry Institute in China discovered a new site for *Cypripedium subtropicum* in Southeast Yunnan, at the border with Vietnam in dense jungle on steep slopes at an altitude of 1,550 m (5,000 ft). Henry Nicholas (Rubber) Ridley, the first director of the Singapore Botanic Gardens, found *Hetaeria oblongifolia* in the Tengah Forest. Thought to be extinct, it was recently rediscovered, 120 years after Ridley first collected it. The species is now being propagated using cuttings made from rhizomes in the garden's orchid nursery. The list goes on.

Odontoglossum crispum Lindl.
Holotype

DET. S. Dalström 20.03

Holotype.
HERB. KEW.

7 What's in a name?

Much work remains to be done to identify and catalogue species. You cannot conserve a species unless you know that it actually exists, and it must have an agreed scientific name and a record of where it is found. Names are therefore important. We need to know that we are all talking about the same thing. In the mid-18th century, Carolus Linnaeus created a binomial naming system for all plant and animal species based on Latin names. Why did he bother? Many orchids have common names in their country of origin, especially if they have some cultural significance, as ornamental or medicinal plants, for example. In Florida and Cuba, *Dendrophylax lindenii* is known as the ghost orchid (*orquídea fantasma* in Cuba), but in Grand Cayman the name ghost orchid refers to a different species, *D. fawcettii*. In the UK, the ghost orchid is *Epipogium aphyllum*. A recently discovered ghost orchid in Madagascar is *Didymoplexis stella-silvae*. Both *Dendrophylax* species are leafless epiphytes, and closely related. *Epipogium aphyllum* and *D. stella-silvae*, however, are terrestrial species, devoid of chlorophyll and completely dependent on their associated mycorrhizal fungi, only appearing above ground to flower.

Sometimes one species may have two or more common names. For example, *Peristylus holochila*, the Hawaiian bog orchid is locally known as *puahala a kane*, stemming from the language of the indigenous people's Polynesian ancestors. Similarly, the official state fish of Hawaii is the reef triggerfish (*Rhinecanthus rectangulus*) but is known by the locals as humuhumunukunukuāpua'a. Clearly, common names don't cut the mustard for the purposes of classification.

When a new species is described for the first time, it is given a unique binomial that scientists throughout the world will recognise, regardless of their native language. The first part of the name refers to the genus to which the organism belongs, which may be made up of many closely related species, or just one. There are a number of different ways of defining a species, but for simplicity, we will say that a species can be loosely defined as 'a group of organisms capable of interbreeding sexually in natural populations that are isolated from other such groups'. Before the advent of DNA sequencing, classical taxonomy depended on measuring and comparing morphological characters. Today's DNA technology has sometimes supported and sometimes disagreed with previous assessments. In either case, it has made the current classification of species more robust.

Initially the budding orchid grower can be put off by scientific names, finding them difficult to pronounce or remember, but familiarity soon resolves the perceived problem. Why Latin? In the past it was the international language of academic discourse, and still has the advantage of being linguistically neutral. Furthermore, Latin is essentially an 'extinct' language and is less vulnerable to change through everyday usage. In any case, most orchids, many

(previous left hand page)
Odontoglossum crispum has undergone a number of name changes since its discovery in the 19th century (and today its accepted name is *Oncidium alexandrae*). The holotype – the single specimen on which the name of the species is attached – is held in the Lindley herbarium at Kew.
Photo: RBG Kew

(previous right hand page)
A large, branched specimen of *Odontoglossum crispum* was exhibited at an orchid show at Jardín Botánico de Bogotá, Colombia.
Photo: Philip Seaton

being tiny and often unremarkable, lack common names, so the only name they have is their given scientific binomial.

So far, so straightforward, but problems can arise, especially for amateur growers, when experts change or modify plant names. A good example is *Cattleya citrina*, which was recently changed to *Encyclia citrina*. Just when growers became accustomed to the change, *E. citrina* then became *Prosthechea citrina*. It is no wonder that name changes can be the subject of much heated debate among scientists and hobbyists alike, especially for popular orchids such as the beloved *Odontoglossum crispum*, recently renamed *Oncidium alexandrae*. But acceptance eventually prevails as we gain a better understanding of where species fit into the evolutionary tree. But name changes certainly don't help the writer when quoting articles from the past!

Taxonomists are scientists who name organisms and attempt to classify them, in other words, they try to figure out where the pieces fit. Taxonomists themselves have sometimes been classified in broad terms as being either lumpers – those who emphasise similarities rather than differences – or splitters – those who emphasise differences rather than similarities (nursery people and collectors tend to like splitting). Why does this matter? Taxonomic decisions can have important consequences in terms of perceived rarity and, with limited funding, prioritising conservation activities. Nor does the neat definition of a species necessarily exist in nature. Reality can be much more complicated, considering that some orchids are extremely promiscuous. In the UK, for example, dactylorhizas form hybrid 'swarms'. Should we focus on conserving individual species, or should we focus on populations?

Applying molecular genetics
We find ourselves at a crossroads where classically trained taxonomists, who seek morphological differences in the species they study, are ageing and retiring, one by one. Like tall rainforest trees, these specialists are succumbing to old age, and will eventually fall leaving behind huge gaps that must be filled, and quickly. Where are the next generation of taxonomists, the seedlings? Unfortunately, in many countries, all too few and far between. Unfortunately, taxonomy is not the academic priority it once was. Will technology come to our rescue?

At an orchid conference in 2007, tropical ecologist Dan Janzen predicted that soon all that would be needed to identify any particular insect species would be a single leg that could be inserted into a hand-held biosensor (conjuring up images of rainforests populated with five-legged arthropods, the sixth limb having been amputated by over-enthusiastic entomologists!). Similarly, the day is not too distant when we will only have to extract a small amount of plant material, and a biosensor will name the species for us. Until recently, an orchid had to be in bloom to be identified with any degree of confidence, and seedlings would be almost impossible to identify. The dream is to confidently identify non-flowering specimens, seedlings, and even pollinia, both in the greenhouse and in the field. DNA barcoding has the potential to turn this dream into reality by providing a way of identifying both plants and animals. Today, college students in the US are using pattern recognition apps on their smartphones to identify life forms such as tiny weeds and drab moths to genus level. While this method is often accurate, it is still not perfect.

DNA barcoding techniques are transforming orchid population studies by allowing scientists to examine the variation within orchid populations. Barcoding compares short, standardised fragments of DNA that are sufficiently diverse to enable us to distinguish between species. All that is needed is a small piece of tissue – a snippet of leaf about a centimetre long is sufficient. To preserve the DNA, tissue samples are dried with silica gel, and kept in a refrigerator or freezer until they are ready to be analysed. Extracting DNA from cells is surprisingly easy. It is routinely performed by teenagers in schools around the world – strawberries are ideal. To break open the tough cellulose cell walls and release the DNA, strawberries are mashed to a pulp in a clear plastic bag to which has been added a soapy, salty solution. After filtering the mush through coffee filter paper, ice cold ethanol is carefully added and, *voilà*, long strands of sticky DNA will appear at the interface of the two liquids.

A more sophisticated technique is needed when using small amounts of tissue, and to obtain a much purer sample of DNA. The entire orchid collection at Atlanta Botanical Garden in the US is currently being assessed, exemplifying the importance and effectiveness of this new technology. The ability to accurately identify plants in the collection using the garden's DNA library is not only providing a better understanding of the plants in the living collection, but also a sound basis for further research. The purified samples are sent to the Center for Biodiversity Genomics in Canada, and the information shared with BOLD (the Barcode of Life Database). Voucher specimens are deposited in the herbarium at Columbus State University and collated with species images. The potential of bar coding for monitoring the trade in controlled orchid species is enormous (see 'Illegal collection', page 144).

Molecular techniques have also been used to better understand orchid mycorrhizal associations. Fungal DNA can now be extracted directly from roots and amplified for analysis in a quick and accurate manner using cost-effective kits that employ PCR (polymerase chain reaction) techniques. This has allowed researchers to identify a wide range of fungi in roots, many to species level. Most companies will also sequence amplified DNA at reasonable cost, saving time and alleviating the need for researchers to purchase their own sequencing equipment. PCR techniques have also been used to compare fungi in pure culture with those present in root samples. Thus, we can now identify the specific kinds of fungi that orchids utilise in their environment to survive.

Genome sequencing

Genome sequencing allows for large stretches of DNA to be analysed rapidly and cheaply, revealing the genetic information that is encoded within a specific region. It has been used to provide a phylogenetic framework or 'tree' for comparing groups of organisms and their evolutionary relationships. In other words, not only can individual species be easily distinguished from one another, they can also be placed onto a family tree of sorts for broader genetic and evolutionary comparisons. When applied to mycorrhizal fungi, genome sequencing can identify a specific strain of a fungus that is most effective at germinating orchid seeds within a cluster of closely related fungi. Prior to this new technology, mycologists spent many hours peering through the eyepiece of a light microscope trying to decipher fungal branching patterns, cell sizes and shapes, and other clues that distinguished one species from another. At best, it was a tedious process that did not always produce reliable results.

VIEWPOINT

Microscopist's thumb
by Philip Seaton

'Don Wimber is working upstairs in the cytology lab. Would you like to meet him?' At the time, Don was the recognised authority on orchid chromosomes, and was taking a one-year sabbatical from the University of Oregon. I was visiting Kew's Jodrell Laboratory for the day. Although DNA studies have transformed our understanding of orchid biology, older techniques such as chromosome staining still have an important part to play. Chromosome number often varies between species within a genus and can be an important distinguishing character. Orchids are usually 'diploid', having two sets of chromosomes, one from each parent – occasionally, however, they have more. For example, some populations of dactylorhizas have four sets of chromosomes – they are 'tetraploids'.

Many who studied biology in school will remember attempting to stain chromosomes in onion root tips. They were placed in a fixative solution to quickly kill the cells and stop cell division. To obtain a layer of cells that was just one cell thick, and after warming them in hydrochloric acid to break down the 'glue' between the cell walls, the tips were gently mushed. Chromosomes were stained with a chemical such as aceto-orcein, and a thin cover slip placed over the top. It was difficult to believe that pressing down as hard as possible with your thumb wouldn't break the cover slip – but it didn't – usually! Peering down the microscope's eyepiece, students were then able to observe cells in various stages of division. Why specifically onions? Because they have eight pairs of relatively large chromosomes. On the other hand, orchid chromosomes are typically tiny and, as I was to discover, *Paphiopedilum philippinense* has 13 pairs of minute chromosomes. By comparison, *Homo sapiens* has 23 pairs of tiny chromosomes, and some ferns have over one thousand.

To say that I was delighted when Don

offered to teach me how to stain orchid chromosomes is an understatement! A few weeks later, I found myself early one morning in Kew's Living Collection with Don searching for suitable root tips, snipping them off, and dropping them immediately into vials of fixative. Early in the morning is best, because that is when many root tip cells are dividing actively (chromosomes can only be seen when the cells are dividing). Back in the lab, the staining technique was essentially the same as I had been teaching my students, but with the addition of those important little wrinkles that somehow never seem to find their way into textbooks. Time, patience, extreme cleanliness, and attention to detail eventually brought their reward in the form of a few precious images in which the chromosomes were nicely spread, and I was able to count them. My thumb was a little sore – but it was worth it!

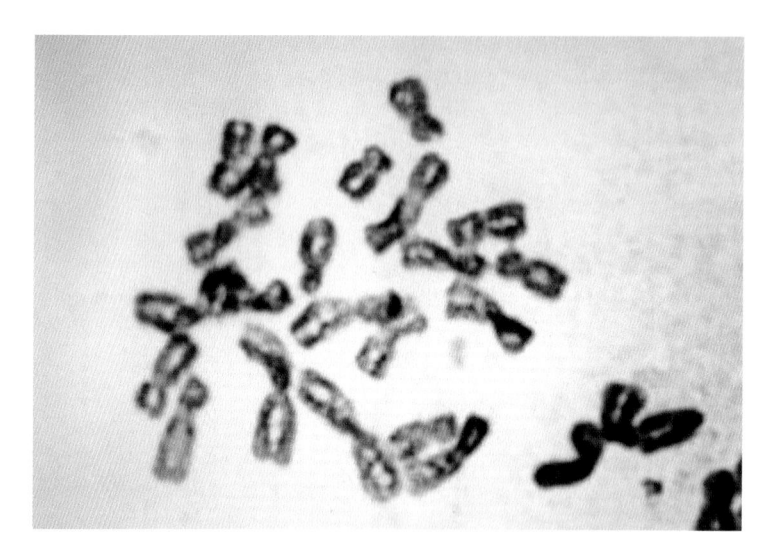

Karyotype of the slipper orchid *Paphiopedilum philippinense*. 2N = 26.
Photo: Philip Seaton

The herbarium

'When I look at a herbarium sheet sometimes all I can see is a dry and dusty specimen. But Diego tells me, 'Tranquilo Guillaume.' He knows and loves his native plants and can often see the plant in its natural habitat in his mind's eye.' Guillaume Gigot talking about his friend Diego Bogarín

In Kew's orchid herbarium, more visitors ask to see *Chloraea magellanica* than any other species. If you close your eyes, you can imagine a young Darwin collecting the plant on Elizabeth Island in the Straits of Magellan on the icy-cold southernmost tip of South America in 1834. The red borders on the paper signify that this is the 'type specimen' – the single physical example (or illustration) of an organism designated as such when the taxon was formally described.

There is one archaic task that the modern, technologically reliant botanist can neither avoid nor escape – preparing and using

> Where flowers are too fleshy to press successfully, they are preserved in liquid (typically 70% alcohol, with 1 or 2% glycerol to prevent the material becoming brittle) in glass jars in a 'spirit collection'. The herbarium at Kew houses one of the world's largest and most diverse collections of orchids in spirit. Unfortunately, the preservative removes the original colour from the flowers, which become pale, straw-coloured mummies in their jars. Both the dried and pickled specimens are supplemented by drawings and colour notes, together with the name of the collector, and the place and date of collection.

herbarium specimens. The word 'taxonomy' conjures up images of crispy brown relics of plants, pressed between ancient pages in a dry and dusty herbarium with the respectful hush of a library, the taxonomists being almost as ancient as their specimens. Yet nothing could be further from the truth. Taxonomy is a dynamic subject, with each new generation committed to building on knowledge of the past and cataloguing the planet's vast orchid biodiversity before it disappears. There is a lot more information contained on a herbarium sheet than the dried plant specimen. Herbaria are therefore important repositories of historical information about where plants were found growing in the past.

As long as the specimens themselves are well cared for, a permanent record will remain of what we have, what we have lost, and where these plants thrived before the world warmed. One problem for botanical gardens in the tropics has been that most of the type specimens for their native orchids are held in herbaria outside of their home countries, limiting the ability of researchers in the tropics to access important specimens. With modern scanners, however, it is now easy to make digital copies of specimens. Digitisation opens up a treasure trove of data – previously only available to specialists through on-site visits or loans – making the information available to the rest of the world via the web. It seems ironic that technology has now seeped into the herbarium cabinet and onto the pages of specimens that are hundreds of years old, giving young (and older) specialists access to knowledge originating hundreds of kilometres away.

A tale of taxonomic confusion

Sitting drinking coffee in a cafe in Mexico City with Phil in 2000, Miguel Soto said: 'Being a taxonomist is like being a detective.' What did he mean? The story of *Cattleya quadricolor* is a perfect example of how the misnaming of species can still cause confusion more than 100 years later.

Opening the cream cabinet door in the Lindley herbarium at Kew, taxonomist André Schuiteman carefully removed the buff folders containing the herbarium sheets. The folder labelled *Cattleya quadricolor* contained two squashed, crispy, toffee-coloured flowers that the botanist John Lindley had received more than a century and a half ago, attached with thin strips of white tape to the herbarium sheet, slightly foxed along its edges. A note in ink between the flowers in Lindley's rounded handwriting read: 'C. quadricolor Rucker Feb 49.'

Holotype of *Chloraea magellanica* in the orchid herbarium at Kew.
Photo: RBG Kew

Cattleya quadricolor first appeared in Europe in 1848, when a plant was acquired by the English orchid grower Sigismund Rucker, who sent flowers to Lindley when it bloomed the following year. On a rectangle of white paper pasted on the bottom left, someone had written in black ink, and underlined, '*Cattleya quadricolor* of Bateman', beneath a printed heading half in red and half in black saying 'holotype'- in other words this is the actual specimen first described as the original model of that species. It seems

104   SAVING ORCHIDS

straightforward enough. Problems arise, however, when a second, different, name is attached to the same species by someone else.

The name *Cattleya quadricolor* first appeared in print in Paxton's Flower Garden (1850–1). In 1864, James Bateman published a description of the plant in *The Gardeners' Chronicle and Agricultural Gazette* under the heading 'New Plants'. He said that *C. quadricolor* was 'distinct from every previously known species of the genus'. He wrote that the petals and sepals were pure white, with the purple, yellow, and lilac of the lip making up the four colours of its name. He also said (wrongly it turns out) that it had been collected in the Magdalena Valley in the Colombian Andes.

In 1870, Linden and Eduard André described a new species, which had been introduced by Linden via his collector Gustav Wallis from the Chocó, a province in Colombia between the Cauca River and the Pacific. They named it *Cattleya chocoensis*. This was followed by a more complete description of *C. chocoensis* in *L'Illustration Horticole* of 1873, accompanied by a beautiful chromolithograph by Pieter de Pannemaeker.

Before we go any further, a word about the geography of Colombia might be helpful. Colombia has two great rivers that run roughly south to north, separated by the barrier formed by the Central Cordillera of the Andes. The Cauca River is on the western Pacific side, while the Magdalena River is located east of the Central Cordillera. Writing in *L'Orchidophile* in 1883, Benedikt Roezl couldn't understand why the name *Cattleya chocoensis* was still being used, as the orchid originated in the Cauca Valley, and nowhere near the region of Chocó. He felt that *C. trianae* or *C. caucaensis* would be more legitimate names. The confusion continued. In the *Manual of Cultivated Orchids* (1887), Veitch considers *C. chocoensis* to be a sub-variety of *C. trianae* (*C. trianae* does indeed grow in the Magdalena Valley). In 1898, Robert Allen Rolfe, editor of *The Orchid Review*, appeared to have had the last word, when he established at last that *C. quadricolor* (which grows in the Cauca Valley) was indeed the correct name.

Why then, are plants of *C. chocoensis* still occasionally encountered for sale, and does it matter? Well, yes, it certainly does if you are interested in conserving a species, because you need to know exactly what species you have in hand. In this case, we are talking about one species, not two. In the case of *Encyclia phoenicea*, we are talking about two species not one (see page 106).

The story of *Cattleya labiata*

The sensation caused by the 'discovery' of the beautiful *Cattleya labiata* marks the beginning of 'orchid fever'. Sometime between 1816 and 1818, William Swainson sent several unidentified plants that he had collected in Brazil to William Jackson Hooker, who was professor of botany at the University of Glasgow (and later became the first director of Kew), and William Cattley (after whom the

VIEWPOINT

One species or two?
by Philip Seaton

I first encountered the chocolate orchid (*Encyclia phoenicea*) in Cuba, growing out of humus in the dead leaf bases of a palm (*Acoelorraphe wrightii*), known locally as guano prieto. As its common name suggests, the orchid's flowers emit a delicious aroma of chocolate and vanilla that attracts bee pollinators (and hobbyists). Many of the trunks were blackened, the trees having been burned to remove their vicious spines, ahead of felling for use as fence posts. Unfortunately the associated orchids went up in smoke as well.

From the roadside, I could see that some palms had not been burned, and still had living specimens attached. The soil was sandy. There were the spines of the numerous *Opuntia* cacti to avoid. The sun was high in the sky. It was hot. The area was swarming with famished mosquitoes. Let's just say that, to get some close-up photographs, I was eaten alive! *Repelente*? Of course, I had insect repellent, but in the face of hordes of hungry mosquitoes it was ... of limited value.

This first encounter had been in the far west of Cuba, in Pinar del Río province. Some years later, I was taken to Ciénaga de Zapata, a shoe-shaped swampy reserve on the Caribbean coast, famous for its unique birdlife. These included Cuba's national bird, the endemic *tocororo* (the Cuban trogon, *Priotelus temnurus*) and the world's smallest bird, the *zunzuncito* (the bee hummingbird, *Mellisuga helenae*). On this trip I saw *Encyclia phoenicea* in habitat that was much wetter and greener. Taxonomists have since decided that *E. phoenicea* should be divided into two distinct species. The one growing near Cortéz is now *E. brevifolia*, whereas the orchid growing in Ciénaga de Zapata is the true *E. phoenicea*. What is the take home message? When collecting plants or seed (with permission) from the wild, it is important to record exactly where the material came from.

genus *Cattleya* was named). The story goes that when Cattley's box of plants arrived, he was curious about the strange-looking packing material and sought to bring some back to life, and that *C. labiata* was not recognised as having any value until it flowered in his greenhouse. The truth is probably more prosaic. It is inconceivable that Swainson was unaware of the beauty of *C. labiata*, having arrived in Brazil on the eve of its flowering season. He must have known how pleased Cattley (and Hooker) would have been with the plants.

A few months later, the plant had not only sent out new shoots, but had also produced huge, rose-coloured flowers with a trumpet-like labellum. It was, Cattley wrote, 'the most splendid of all Orchidaceous plants, which blossomed for the first time in Britain in the stove of my garden in Suffolk'. A plant subsequently flowered at Glasgow Botanic Gardens in 1819. Eventually, a small number of plants found their way into collections. Joseph Paxton reported that, 'our plant flowered in the Orchideæ House at Chatsworth' in the autumn of 1837, and that the plants were, 'available from Loddiges, Rollison or Knight.'

Meanwhile, Swainson had emigrated to New Zealand, and the source of *Cattleya labiata* was widely assumed to be in the Organ Mountains, north of Rio de Janeiro. This was largely due to the Scottish botanist George Gardner, who wrote that he had visited the 'Tijuca Mountains' in 1836 where, 'On the face of the mountain, at an elevation of several hundred feet, we observed some large patches of those beautiful large-flowered Orchidaceous plants that are so common in Brazil. Its large rose-coloured flowers were very conspicuous, but we could not reach them. A few days afterwards we found it on a neighbouring mountain and ascertained it to be *Cattleya labiata*.' A few days later he climbed the Pedra Bonita. 'On the edge of a precipice on the eastern side, we found, covered with its large rose-coloured flowers, the splendid *Cattleya labiata*, which a few days before we had seen on the Gavea.'

When he visited the same spot the following year, he found that the orchid's natural habitat was rapidly disappearing, to be replaced by coffee and other crops. 'The forest, which formerly covered a considerable portion of the summit, was now cut down and converted into charcoal; and the small shrubs and Vellozias which grew in the exposed portion, had been destroyed by fire. The progress of cultivation is proceeding so rapidly for twenty miles around Rio, that many of the species that still exist, will in the course of a few years, be completely annihilated, and the botanists of future times who visit the country, will look in vain for the plants collected by their predecessors.'

As a result of Gardner's words, many would-be-collectors of this iconic species were sent on a wild goose (well, orchid) chase, urgently scouring the area around Rio de Janeiro in the hope of collecting plants before the remaining forest was completely

Nectar spur of *Dendrophylax lindenii* pictured in Guanahacabibes National Park, Cuba in January 2016. The nectar spurs of Cuban ghost orchids are longer than those in Florida.
Photo: Lawrence Zettler

converted into coffee plantations. All to no avail. Not a single plant was found. It seemed that Gardner's prophecy had come true: the habitat had been destroyed and *Cattleya labiata* was no longer to be found in the wild. In fact, they were looking in the wrong place. It turned out that Swainson had collected *Cattleya labiata* more than 2,000 km (1,200 mi) north of Rio de Janeiro, in the state of Pernambuco, and Gardner had mistaken *Laelia lobata* for *Cattleya labiata*. In a letter to Professor Jameson, Swainson had written: 'Instead of following the example of all my fellow-labourers, by going in the first instance to Rio de Janeiro, I landed, about the end of December, 1816 at Recife, in the Province of Pernambuco. This province has never been visited by a modern naturalist.'

Eventually the home of the plant was rediscovered, and it was exported in large numbers. *Cattleya labiata* can still be found today in Pernambuco. The original native forest has largely been cut down and turned into agricultural land, but in some of the remaining forest fragments, flowering plants can still be found, together with seed capsules and seedlings. Likewise, *Laelia lobata* can still be found growing wild today in a limited area in the vicinity of the city of Rio de Janeiro, growing high on rocks facing the Atlantic Ocean, and fully exposed to the sun.

A painting of *Cattleya labiata* from John Day's Scrapbooks.
Painting: RBG Kew

CATTLEYA LABIATA Var. pallida

Summer Flowering Variety.
C.L. var. pallida   No 5, of Catalogue

Sept 12th 1867

This plant is flowering late this year. This variety generally blooms in July or August.

Flower generally in pairs — sometimes 3 upon a Scape.

It is very delicate & lovely but not so large & showy as the ordinary variety which blooms in Oct & Nov.

Pollinia
n.
Enlarged

Lip spread out.

Margin exceedingly crisped.

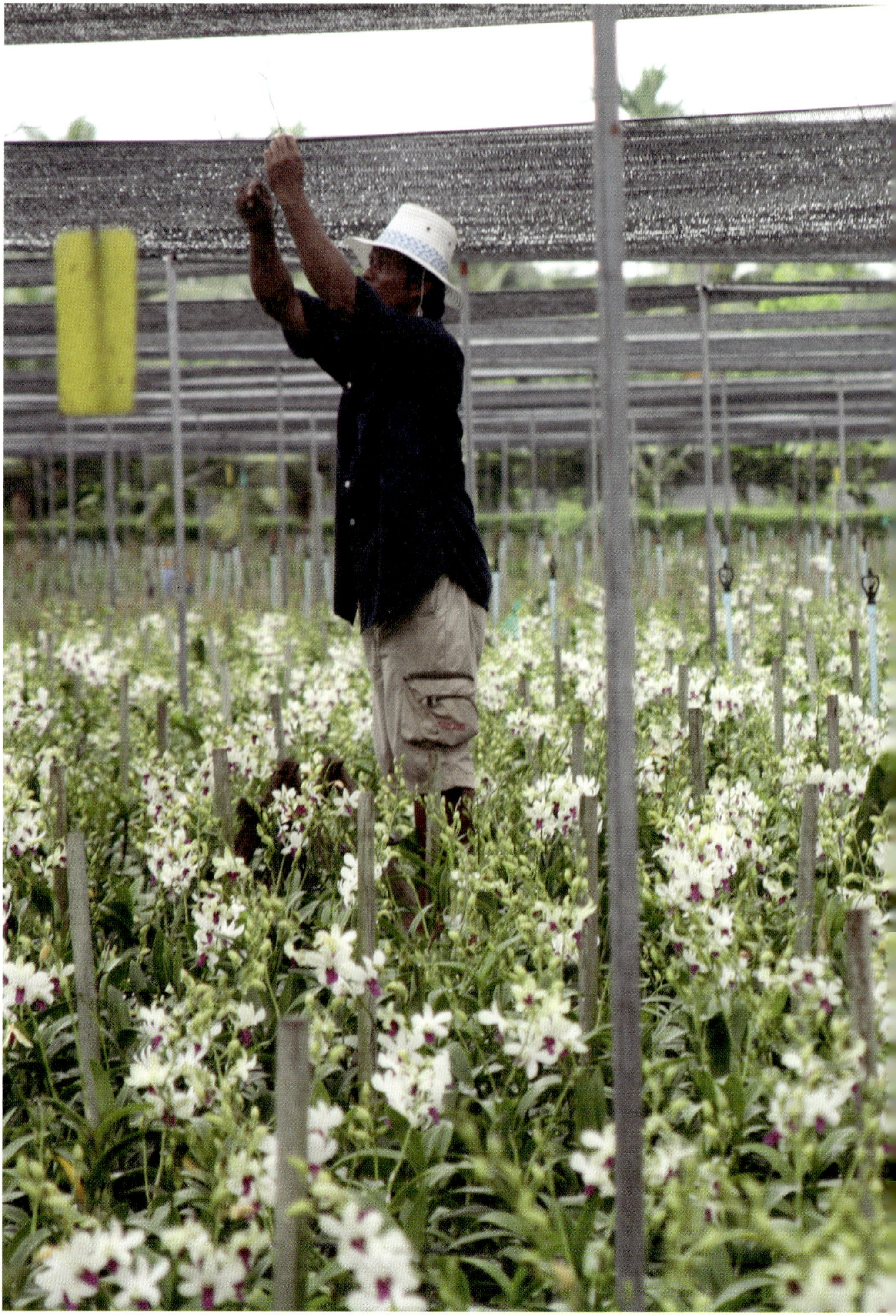

8 Why we should care

Do we need a reason to conserve tigers? Although we get dewy-eyed about the potential loss of a large furry mammal, it seems we have lost the feeling that plants are intrinsically important. Tigers sometimes kill people, after all, and, as far as we are aware, no orchid has done anyone any harm. Do we wish to conserve tigers merely because they are a source of tiger bone, mistakenly thought by some to be a potent aphrodisiac? Of course not. We conserve them because they are part of our natural world, as are plants. Put simply, orchids enrich our lives.

In the same way that roses are loved in England, people in many tropical countries are proud of their native orchids. In Costa Rica, such is the love that *ticos* have for their national flower, *Guarianthe skinneri*, that there are more growing in gardens in that country than remain in the wild. In their wild state, epiphytic orchids are limited by the lifespan of their host tree. Such constraints don't apply to cultivated plants, which often grow into huge specimens. Enormous baskets of *G. skinneri* varieties, for example, can be seen at orchid shows in the Costa Rican capital, San José.

(previous pages)
Harvesting *Dendrobium* flowers in Thailand. Growing orchids for cut flowers is annually a multi-million pound global industry.
Photo: Philip Seaton

(below)
Philip Seaton with *Paphiopedilum insigne* in Madeira.
Photo: Joyce Seaton

The pot plant and cut flower industry

Once upon a time, the crystalline white blooms of the cool-growing *Coelogyne cristata* and the glossy slippers of *Paphiopedilum insigne* were grown in shallow pans in the greenhouses of parks departments around the UK, ready to be used in floral displays at civic events. In the 19th century, sprays of the free-flowering *C. cristata* were popular in wedding bouquets. Today, tourists visiting the island of Madeira early in the New Year will see pots of *P. insigne* everywhere, and their cut flowers are placed in small vases as centrepieces on restaurant tables.

Today, orchids as pot plants are big business. Moth orchids (*Phalaenopsis* hybrids), which can be found for sale in many supermarkets, are a multi-million dollar industry in the Netherlands and Taiwan, and adorn windowsills around the world. They can often be seen in the background of television programmes, where their elegant sprays of flowers are used to add a hint of sophistication. In Singapore and Thailand, dendrobiums, arandas, ascocendas, and vandas are grown in vast quantities for export as cut flowers. Orchid nurseries supply plants to tens of thousands of amateur growers. All depend on plants that were collected from the wild at one time or another, and the remaining wild plants – if collected responsibly and in accordance with local environmental laws – remain a possible source of novel genetic material for future hybrids.

Orchids in culture

Orchids have a long history in Chinese and Japanese culture. The deliciously scented *Cymbidium ensifolium* has been grown in China for more than 2,000 years. In Japan, the cultivation of *Vanda falcata* (syn. *Neofinetia falcata*) goes back more than two centuries to the Edo period, and today there are societies devoted solely to its cultivation. The Japanese refer to it as *fuuran* (wind flower) or the samurai orchid, because in pre-industrial Japan only samurai warriors were allowed to cultivate it. The plants resemble miniature bonsai and are often grown as much for their leaves as for their flowers. For judging at shows, they are displayed in beautiful ceramic pots that complement the foliage. The snow white flowers, with their elegant long pendent nectaries, have a sweet fragrance that has been described as reminiscent of lily of the valley (*Convallaria majalis*), with hints of ripe lemons. Native to Japan, Korea, and China, today it is considered endangered in its forest habitat, where it grows as an epiphyte or, rarely, as a lithophyte.

In Latin America, people often make offerings of flowers, food and drink when they visit the graves of their dead relatives. These occasions frequently involve the use of orchids as decorations. Orchids have special cultural significance in Mexico, in particular. In pre-Columbian Mexico, mucilage was extracted from the dehydrated and ground-up pseudobulbs of laelias and other orchids to make a glue to bind feathers on head-dresses. Local children still play with the beautiful golden flowers of the slipper orchid *Cypripedium irapeanum*. They pop the flowers' pouch-like lips by slapping them between their hands or use them as whistles. In Mexico, Guatemala, and Honduras the pseudobulbs of *Prosthechea michuacana* are used to quench thirst.

Citizens of India have inherited an ancient culture that is tolerant of wildlife and they also value plants for food, spices, and traditional folk remedies. Orchids such as *Aerides, Bulbophyllum, Paphiopedilum* and *Vanda* are especially cherished as a source of fragrances and cut flowers, in addition to their use in traditional medicine. This respect for nature, and the knowledge needed to sustain it, has been passed down from one generation to another. As long as this cycle is repeated, there is hope. The task ahead is to keep the younger generation connected to the natural world.

Medicinal orchids

Medicinal plants remain an important traditional therapeutic resource for many indigenous communities around the world. Unfortunately, the increasing acculturation of such communities

(above)
Vanda (*Neofinetia*) *falcata* displayed in a traditional pot.
Photo: Mike Bull

(opposite)
Prosthechea karwinskii in an oak forest in Oaxaca, Mexico.
Photo: Rodolfo Solano

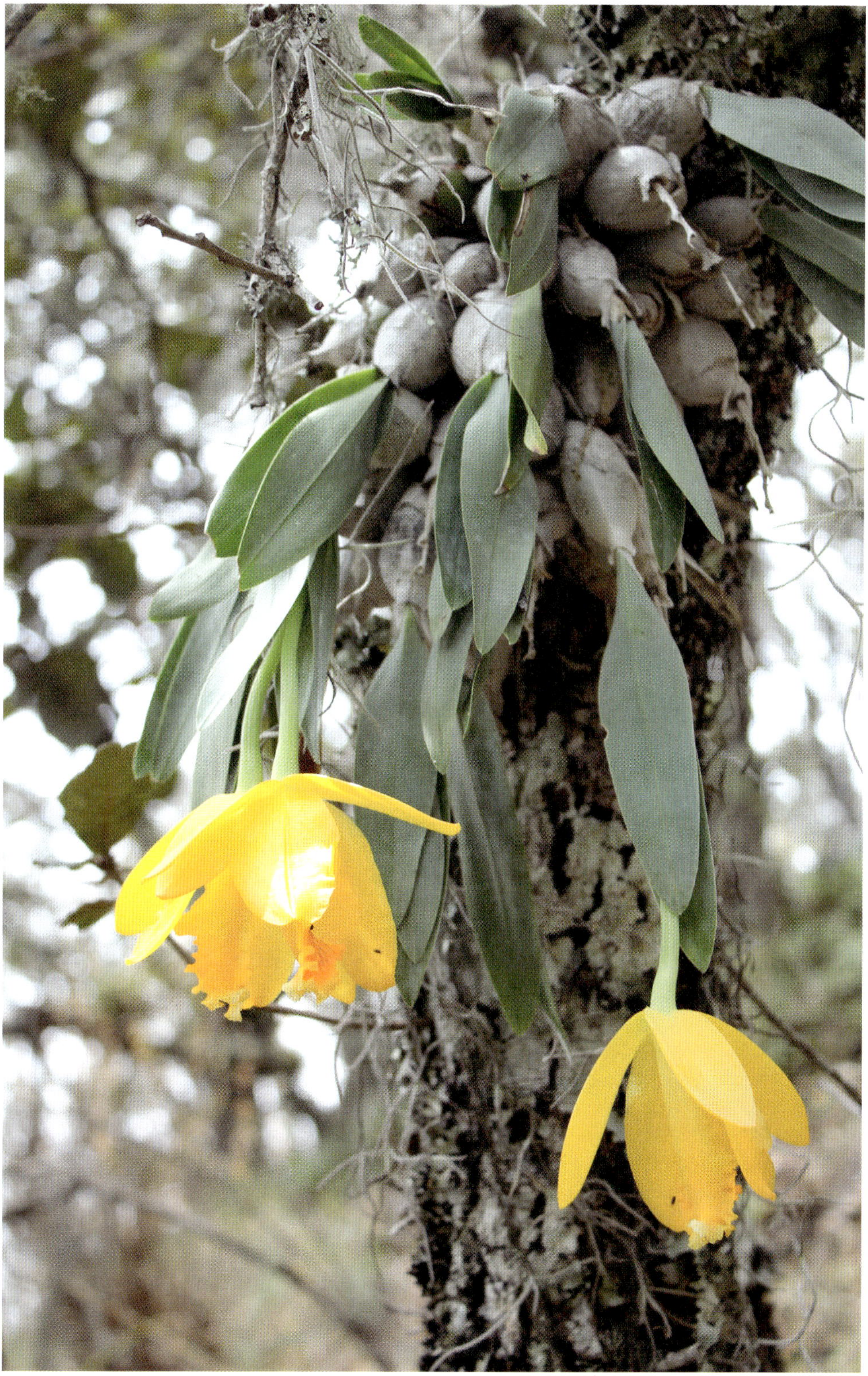

means that much of their knowledge about plants that may contain pharmacologically valuable chemical compounds is in danger of being lost.

Orchids have been, and continue to be, used in traditional medicines. In Europe, the tubers of *Ophrys* were once used as an aphrodisiac, based on their resemblance to testes (the word 'orchid' is derived from the Greek word for testicle). In the central highlands of Peru, tables are decorated with vases of the flowers of *Masdevallia uniflora* and *M. rimarima-alba.* The latter is also used as a traditional medicine, supposedly to help children to begin speaking (*rima rima* means 'speak speak' in Quechua). The flowers are still harvested from the forests, but as the old traditions begin to wane, and the forest is no longer valued, the trees are being progressively cleared for cultivation – principally of potatoes. In Africa, different species of *Eulophia* are used to treat diabetes and epilepsy. Likewise, in Mesoamerica there are numerous examples of orchids that are used in traditional medicine to treat a bewildering range of conditions. The pseudobulbs, leaves, and flowers of *Prosthechea karwinskii* are used to treat a range of conditions: coughs (as infusions), wounds and burns (as poultices), and diabetes (chewed), and to prevent miscarriage and assist in childbirth (as infusions). As *P. karwinskii* blooms around Easter, it is also used in religious decorations in homes and churches.

More than a third of China's orchids are used in traditional Chinese medicine. Many have been used since ancient times. *Dendrobium* species in particular continue to be used as essential ingredients in traditional medicines, and are still harvested from the wild. Unfortunately, local communities sometimes assume that wild-collected plants have superior medicinal qualities compared with those that are cultivated. *Dendrobium moniliforme*, with its sweetly fragrant and compact flowers, is not only used in Chinese medicine but also grown as an ornamental plant in Japan. Tianma, the dried tubers of *Gastrodia elata*, has a long history of being used to treat a variety of disorders, including diseases of the nervous system. Devoid of chlorophyll, *G. elata* is entirely dependent on sugars produced by the honey fungus (*Armillaria mellea*). An international centre is being established to stimulate research into the orchid biodiversity in the Yachang Orchid Nature Reserve. One of the first objectives is to raise species such as *Dendrobium officinale* from seed for subsequent cultivation and sale by local farmers.

In Nepal, 90 orchid species have been reported to have medicinal properties, and the demand for illegally collected orchids in that country and region is a growing problem that puts the remaining populations at serious risk of extirpation. The Nepalese botanist Bijaya Pant believes there will be a large demand for her country's orchids in the future. This can be sustainably met through the cost-efficient application of *ex situ* tissue propagation techniques. The hope is that people will realise the long-term

Propagating medicinal orchids in a laboratory in Yachang, China.
Photo: Philip Seaton

value of orchids in generating income, promoting conservation, and facilitating ecotourism. Pant's team, in association with the Annapurna Research Centre in Kathmandu, have begun mass producing native orchids, which have both aesthetic and medicinal value. They are encouraging growers as well as the wider community to cultivate these orchids as a means to minimise illegal collection and trade. As a result, people are increasingly interested in propagating orchids produced by tissue culture, and have begun growing the seedlings. The team are planning to continue this programme in association with various partners on a larger scale.

Flavouring and perfumes

Have you ever wondered where that little brown bottle of vanilla extract at the back of your kitchen cupboard came from? People are often surprised to learn that the long, thin black vanilla 'beans' sold in individual sealed tubes in supermarkets are actually seed capsules of an orchid, mostly *Vanilla planifolia*, which are also used to make vanilla extract. Originating in the forests of Mexico, the Aztecs were the first to use vanilla 'beans' (*tlilxochitl*) as a flavouring for their traditional cocoa drink (chocolate). Today, vanilla is the world's favourite ice-cream flavour. The black flecks in the luxury brands are the tiny seeds. The vines are grown either under shade cloth and trained up wooden frames, or in small trees whose canopies provide shade. The flowers open at dawn and last for only one day. In the early 1800s, the French introduced

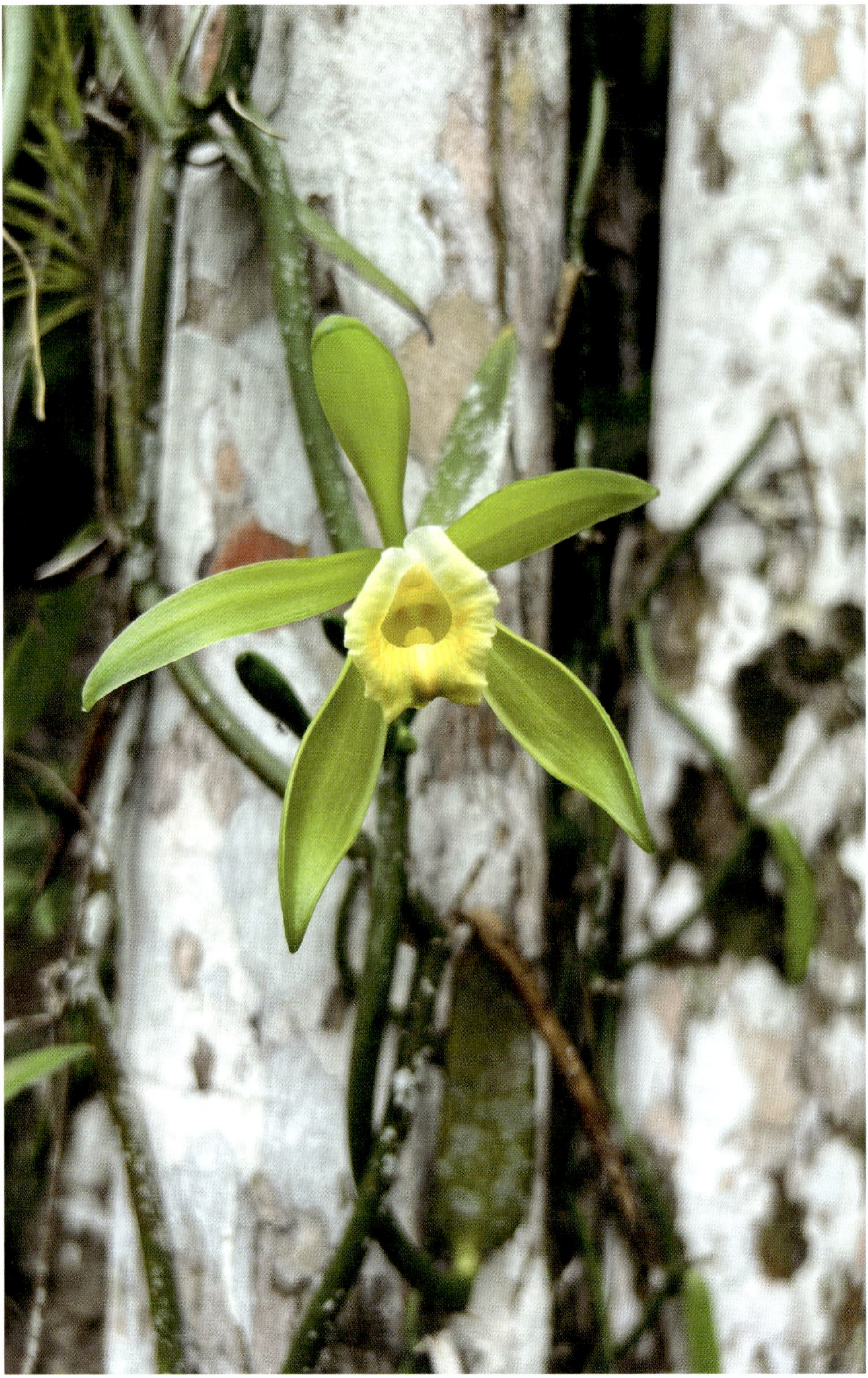

vanilla to Réunion Island and Madagascar. Today, Madagascar is the world's largest producer but, because there are no native pollinators on that island or indeed anywhere in the Old World, the flowers must be hand-pollinated in the early morning.

Because it is easily propagated from stem cuttings, vanilla 'farms' or vanilariums consist of just one clone. This narrow genetic base makes commercial vanilla cultivation vulnerable. An epidemic of a new strain of one of the current fungal diseases, for example, has the potential to devastate the crop in the same way that Panama disease (*Fusarium* wilt) poses a global threat to the Cavendish banana, and coffee leaf rust continues to devastate plantations of *Coffea arabica*. This lack of genetic diversity in our monoculture world is further exacerbated by the lack of diversity in wild plants, and there is an urgent need to find new clones. *Vanilla planifolia* is not the only species that is appreciated for its flavour and perfume. There are around 100 known *Vanilla* species, some of which are (illegally) collected from the wild for sale in the markets of cities such as Medellín and Bogotá in Colombia.

Eating to extinction?
The orchids that are used as a food source are by no means limited to *Vanilla*. In 17th century England, salep – made from orchid tubers with a gummy texture – was a popular beverage sold on the streets of London, until it was replaced by tea and coffee. Still popular in some Mediterranean countries, primarily Turkey, salep is made by grinding the dried tubers into a powder to make hot drinks, ice cream, and other desserts. In India and Pakistan, it is sold as an aphrodisiac. Millions of orchid tubers are harvested from the wild each year, largely in Turkey and Iran. The trade affects as many as nineteen orchid species, and many local populations are threatened with extinction. What can be done? Perhaps the story of chikanda can provide some clues.

Chikanda or African polony – a cake with a meat-like consistency made from ground orchid tubers and peanut flour – is a traditional food in Zambia. Originally consumed as a local delicacy, today it is sold as a luxury item in hotels, restaurants, and supermarkets. Increasing demand and unsustainable harvesting have led to the decimation of many terrestrial orchid populations. In 2014, TRAFFIC (a non-government organisation that monitors the illegal trade in wild plants and animals) found that, in addition to the millions of tubers being harvested in Zambia, tubers were being imported from the surrounding countries. Women and girls in rural communities were harvesting the tubers to supplement their incomes, and they were walking increasingly long distances over many days to obtain them. The practice was clearly unsustainable, and in 2016 a Darwin Initiative project (funded by the Department for Environment & Rural Affairs of the UK government) was established by Kew and Copperbelt

Vanilla planifolia photographed growing in Cuba.
Photo: Lydia Ballard

University in Kitwe, northern Zambia. Led by Ruth Bone at Kew, and coordinated by Nicholas Wightman in Kitwe, the long-term aim was to enable members of the rural population to cultivate the orchids, giving them a reliable source of income, while reducing pressure on declining natural populations.

Conservation often brings together people and their expertise from many different parts of the world. In this case, a series of workshops were delivered by Serene Hargreaves (Kew) focusing on assessing the plants' conservation status; Jonathon Kendon (Kew) taught participants how to isolate and culture mycorrhizal fungi; Hildegard Crous (Cape Institute of Micropropagation, South Africa) used her experience with *Disa* orchids to demonstrate both immature (green pod or capsule) culture and asymbiotic germination techniques; and Phil Seaton (Orchid Seed Stores for Sustainable Use) advised on orchid seed collection and storage, germination, and viability testing. To establish exactly which species were being utilised and to assess their conservation status, Bone had collected samples for identification in the field. DNA voucher specimens were deposited both in the herbarium at Kew and in the Division of Forest Research Herbarium in Kitwe. DNA barcoding techniques revealed that 16 species in six genera were being harvested, including *Disa robusta*, *Platycoryne crocea*, and *Satyrium buchananii*.

Nicholas Wightman bought tubers of unidentified orchids at local markets to cultivate in a purpose-built shade house on his farm, and efforts are currently under way to harvest the species in a more sustainable manner. One notable success of the project has been the development of the Chikanda Orchid Key – an orchid identification app (Google Play) used to identify orchid species for the purposes of hand pollination to generate seed.

Felix Chileshe from the Copperbelt University and a roadside chikanda seller in Zambia.
Photo: Jonathan Kendon

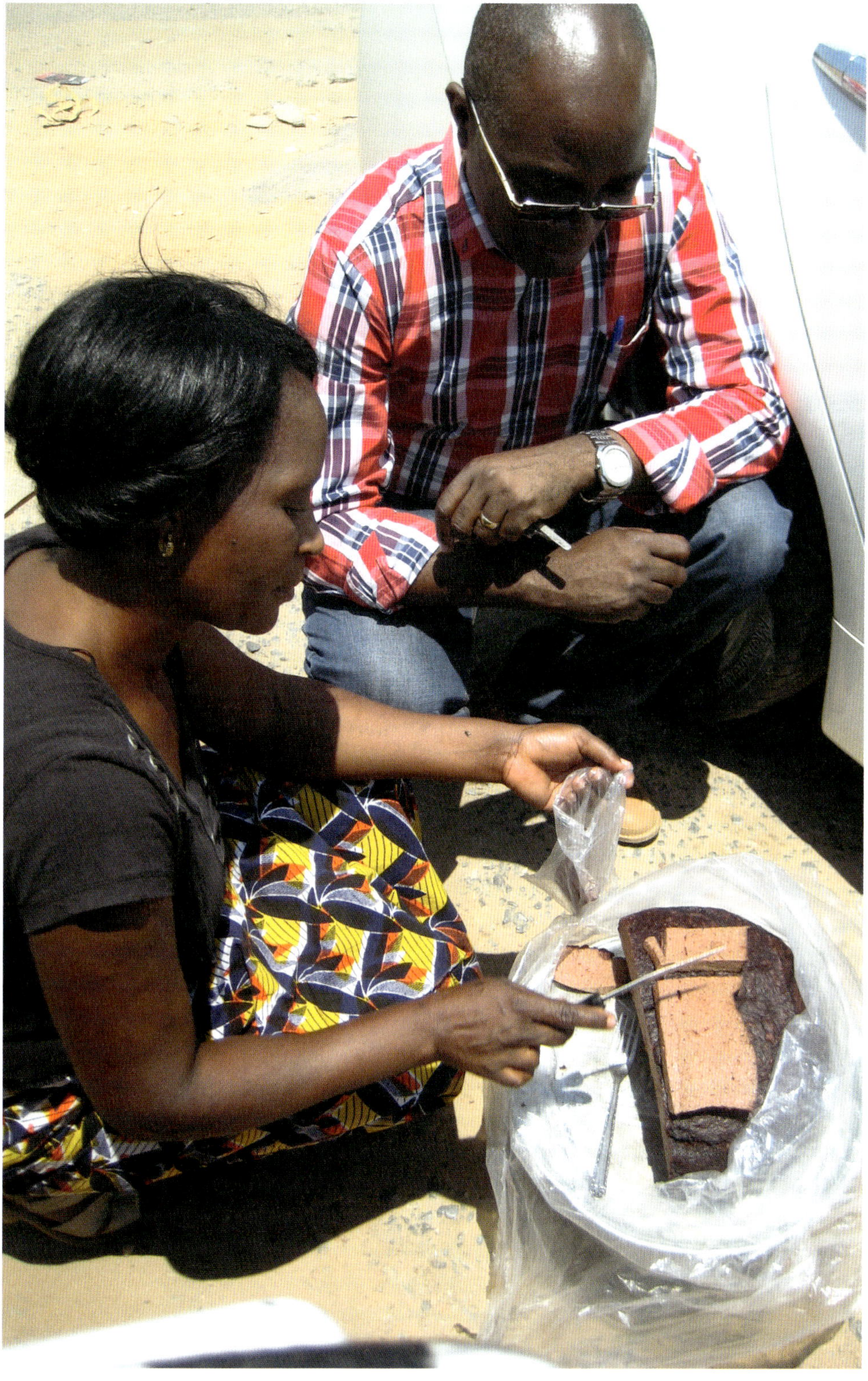

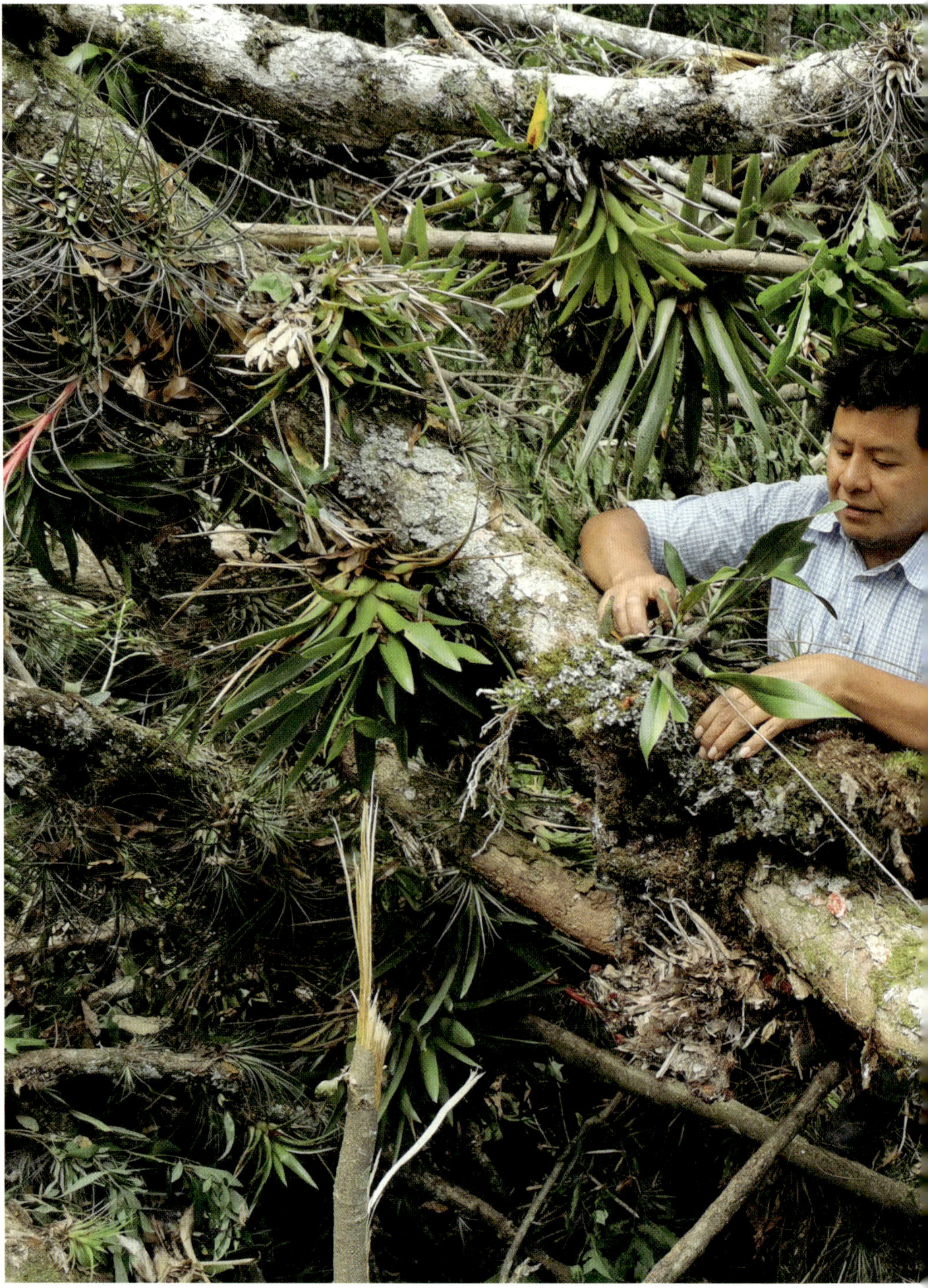

9 Life (and death) in the anthropocene

Extinctions

'It is sometimes frustrating, when discussing the effects of human activities on the conservation of orchids in a country like Costa Rica, which in less than two centuries has lost more than two-thirds of its forests, that we cannot cite a single concrete case of a species indisputably extinct. Based on our knowledge of the very strict local distribution of some taxa, which may not exceed a single hill or valley, it is rational to suppose that a large number of orchid species must have disappeared from the hundreds of 'unique' valleys and hills and mountains that have been completely barren of any original vegetation. However, the evidence necessary to transform these suppositions into scientific facts is lacking.' Franco Pupulin, 2014

On the basis of habitat loss alone one would expect that the world is haemorrhaging orchid species. Yet we haven't really got a handle on which, and how many, have gone extinct. Thirty years ago, Harold Koopowitz suggested that, 'With 45% of the tropical forests already cut we should have lost approximately 22% of all orchid species' and that, 'as many as 402 of the 3,405 pleurothallid species may already have been lost.' According to Kew's 2020 report *Status of the World's Plants and Fungi*, as many as two out of every five species of vascular plants and fungi may be threatened with extinction. If true, this would mean that around 12,000 orchid species are currently at risk.

Deciding when a species has become extinct in the wild is fraught with difficulty because, however unlikely, there often remains the possibility of discovering another population. The story of *Paphiopedilum delenatii* gives pause for thought. First found in Vietnam between 1913 and 1914, the French explorer Poilane collected several plants in 1922. All of them died, apart from a single plant that was cultivated in the nursery of Vacherot and Lecoufle in France. Thought to be extinct in the wild for many years, all of the plants in cultivation were derived from self-pollinating this lone plant until, in 1992, wild-collected specimens appeared for sale in Taiwan after the species' rediscovery by local people. Likewise, *Paphiopedilum druryi* was assumed to be extinct in the wild for many years, but has since been rediscovered, with over 3,000 specimens known from one extant population. Ironically, dreadful summer fires along Australia's east coast in 2020 led to the rediscovery of the lemon scented 'mignonette leek orchid', *Prasophyllum morganii*. First documented in 1929, the orchid had not been seen since the 1930s. Researchers assessing the damage to wildlife in the burnt areas of New South Wales have since found four populations.

We must be clear about what we mean by the term 'extinct'. The term 'extirpated' is sometimes more appropriate, meaning that a species may have disappeared from one region, but it might remain somewhere else. For example, although extinct in the

(previous pages)
Agustin Ramos, on a short expedition in central Veracruz, Mexico to monitor felled trees and see whether orchids could be found in them. Logging for firewood and construction is still very common in small communities in this area, and often orchids are unknowingly destroyed. This tree is a species of *Liquidambar*. The orchid pictured in the middle of the image is *Prosthechea vitellina*.
Photo: Elias Ramos

wild, the Mexican species *Laelia gouldiana* remains common in cultivation. While it is difficult to name examples of orchids that have become globally extinct, it is not difficult to find species that will soon meet that fate without human intervention. We should also consider the issue of rarity and abundance. Orchids may be naturally rare due to having small populations, narrow ranges, precise niche habitat requirements, and symbiotic relations with specific fungi and pollinators. The Mexican species *Mexipedium xerophyticum* and *Phragmipedium exstaminodium* subspecies *exstaminodium*, for example, are extremely vulnerable as their populations are so small, and their habitats have either disappeared or are in danger of disappearing.

Rather than focusing exclusively on species (and whether two or more geographically isolated populations are distinct species), we should perhaps focus more on populations. It is not just the number of species in decline, but also the number of populations, and the sizes of those populations – that is to say, the numbers of individual plants. For example, *Dendrophylax lindenii* occurs both on the Guanahacabibes Peninsula in Cuba and in the Panther Refuge in southern Florida, separated by around 600 km (373 mi), and by water. We cannot imagine either country being happy to allow their own population to be extirpated on the basis that it can always be found in the other place!

Red Listing

How do we know which species are endangered, in other words those that require urgent conservation action? The Red List of Threatened Species was established by the IUCN as an authoritative and comprehensive source of information to address this question. After application of a series of standardised criteria, species are recorded as belonging to one of five categories: Extinct, Extinct in the Wild, Critically Endangered, Vulnerable, of Least Concern and Data Deficient. Red Listing has been described as a barometer of life. Regular updating of the assessments, and comparing them with earlier results, enables conservationists to determine whether a species has become more or less endangered. Assessing the conservation status of perhaps 30,000 species is a daunting task. As of September 2020, 1,641 orchid species had been assessed. Currently 197 species are listed as Critically Endangered, 355 Endangered, and 195 Vulnerable, but only five extinctions have been documented on the Global Red List. By the time this book is published, doubtless more species will have been assessed, but clearly a lot remains to be done.

In addition to the Global Red List, many countries maintain their own lists of endangered orchids. The distinction between the two types of lists is important, because what might be considered endangered in one country may be common in a neighbouring country. For instance, the sword-leaved helleborine, C*ephalanthera* *longifolia*, is rare in England, but is common in many European countries. Do we want to maintain populations of *C. longifolia* in the UK? Of course we do! In other words, the Global Red List and country lists are complementary.

The stories of *Paphiopedilum fairrieanum* and *Kefersteinia retanae* illustrate the problems in determining the status of orchids in the wild. When *P. fairrieanum* first appeared for sale at Stevens' sale rooms in London in 1857, the source of the plants was unknown, although it was known to be somewhere in Assam (northern India) or Bhutan. By 1881 the species was already rare in cultivation, and by 1904 it was said that only a single plant remained in cultivation in England. If ever an orchid could be said to be full of personality, it would be *P. fairrieanum*. With a dorsal sepal marked with bold crimson-purple veins and lateral sepals resembling a handlebar moustache, it is unique in the world of slipper orchids. Such was its desirability that in December 1904, the nursery Messrs Sander offered a reward of £1,000 for its rediscovery. Understandably, this created a lot of interest (and publicity for the nursery). In 1905 the plant was discovered by a Mr Searight in Bhutan, who claimed the prize. Whether or not he received all the money remains unknown.

Fifty years ago, wild-collected plants of *Paphiopedilum fairrieanum* could still be seen for sale in substantial numbers in at least one commercial orchid nursery in the UK. With a reputation as being difficult to grow, however, *P. fairrieanum* became rare in cultivation in the UK once more, and was seldom seen in orchid show displays. There is some good news, however. Such is the desirability of this orchid gem, that seed-raised plants have recently become widely available, demonstrating that orchids that are on the brink of disappearing from cultivation can be rescued where there is a will to do so.

According to the IUCN website, it was assessed in 2014 as being Critically Endangered as it was very local, with only 49 mature individuals remaining in a restricted area in India. As the population was decreasing due to threats, including ruthless collection for the horticultural trade, habitat degradation, human disturbance, trampling by cattle, deforestation, and forest fires, it was thought to be on the verge of extinction in the wild in India, and probably extinct in Bhutan.

But was this orchid really Critically Endangered? Stig Dalström has been working in Bhutan as part of the Thunder Dragon Orchid Team, together with forest rangers and local men in their 70s who remember the orchids being collected more than 50 years ago. The team has found many new populations, some of them consisting of a thousand plants or more. Dalström says: 'It turns out that once you know where to look this species turns out to be pretty abundant today.' The team now considers the species to be safe in its natural habitat. Since Dalström's report, an article has appeared

Cultivated plant of *Paphiopedilum fairrieanum*.
Photo: Joyce Seaton

127   LIFE (AND DEATH) IN THE ANTHROPOCENE

128 SAVING ORCHIDS

in *Orchids* magazine written by Udai and Hemlata Pradhan, describing a trip to a location close to the border with Bhutan in 2012, where they also saw healthy populations of the orchid.

In 2021, Susanne Masters and Diego Bogarín reported that in Costa Rica, 'a remnant of forest in the El General valley, measuring just 200 by 100 m (660 by 330 ft), that had escaped conversion to coffee plantation, hosted a small population of *Kefersteinia retanae*.' Described in 1995, and known in the wild from this single locality, a 2003 survey recorded 22 plants, including some that were fruiting. A subsequent survey raised that total to 30, with a few extra known to be in cultivation. In 2013, the orchid enthusiast who discovered the species, Jorge Cambronero, visited the site and saw that all the *K. retanae* had been collected illegally. At the time when Cambronero was writing, the orchid was thought to be extinct in the wild and persisted as just a few individuals held in private collections. While habitat loss due to agriculture had pushed *K. retanae* to the brink of extinction in the wild, illegal collection administered the hammer blow. A few months later, Bogarín reported that, 'a friend just told me that he has found *Kefersteinia retanae* in Osa Peninsula close to Parque Nacional Corcovado, so this species should be around! I am really happy that it was found in another locality.' Thus, the stories of *Paphiopedilum fairrieanum* and *K. retanae* both demonstrate the difficulties of determining the conservation status of many species, while at the same time providing hope that some of the 'lost species' are still out there.

Are orchids being safeguarded in reserves?
For each country, we need to know how many species there are, which orchid species are conserved within protected areas such as national parks and private reserves, and how many are found outside such reserves. For instance, we know that a large proportion of Mexican orchids occur outside protected areas. A study conducted in Taiwan showed that large areas of high orchid richness were excluded from existing conservation reserves, and were insufficiently investigated. Despite its subtropical location and being surrounded by warm water from the South China Sea, the island's forests have experienced periods of drought in recent years linked to climate change. Rebecca Hsu tells us that many of Taiwan's 400 species occur in remote regions that are difficult to sample, and information about their distribution is limited. This lack of knowledge about distributions hampers conservation planning for extinction-prone species, and so modelling techniques may be used to identify potential refuges for rare species, and to evaluate the effectiveness of the present reserves.

Colombia and Ecuador, which are found in the richest biological 'hotspot' on Earth, are a different kettle of fish. With more than 4,000 apiece, both neighbouring countries seek the distinction of harbouring the highest number of orchid species.

Kefersteinia retanae, an endangered species in Costa Rica.
Photo: Diego Bogarín

Ecuador is home to 4,300 orchid species, and counting, over half of which have been described within the past 50 years. Roughly 60% (2,480 species) are restricted to the Andean mountains between 1,200 and 4,200 m (3,900 and 13,780 ft), which makes the region difficult for loggers to access. Around 620 of these species are endemics that are restricted to small geographic areas. Plagued by the history of political unrest in these two countries, especially in Colombia, many new species await discovery in remote regions deemed unsafe for travel. Meanwhile, high levels of poverty fuel the appetite for the clearing of forests for locals to plant crops. As a result, many new orchid species may become extinct before they are discovered and described.

Extending from a northern temperate zone, south to the northern edge of the tropical zone, with enormous variations in climate and topography, China harbours another plant biodiversity hotspot. Much of its biodiversity has been lost forever, however, and climate change is having a negative impact on many of the

Paphiopedilum hirsutissimum growing in the Yachang Reserve, Guangxi, China.
Photo: Holger Perner

remaining orchid populations (see Holger Perner's observations in 'Drought, fire, and floods', page 155). Nevertheless, there are bright spots that remind us of what the world was once like. Despite the depredations of the past (and present), some sizable populations of orchids do remain. Yachang, in Guangxi Province, lies at the border with Guizhou province, close to the Yunnan border. The area (Leye County) is called 'little Manchuria' because it has the coldest winter in South China, with a minimum temperature of -2°C (28°F). In addition to an estimated 100,000 individual plants of *Paphiopedilum hirsutissimum* and 12,000 *Cymbidium cyperifolium*, sub-tropical Yachang is home to 120 orchid species in 107 genera. *Coelogyne fimbriata, Eria rhomboidalis, Panisea cavaleriei Bulbophyllum kwangtungense* and other species cover exposed rock faces in large colonies. The density of some species can reach 300 plants per square metre. However, orchids in the area remain threatened by habitat destruction and illegal collection.

A century after Ernest Wilson's journey through Western Sichuan in China, Mark Flanagan and Tony Kirkham from Kew retraced his steps in an attempt to photograph the same places that Wilson had visited, and to discover how the country had changed. In June 1908, Wilson had described the flora of a grassy ridge leading up to the Pan-lan shan Pass. 'All the moorland areas are covered with the Thibetan [sic] lady-slipper orchid (*Cypripedium tibeticum*) so that it was impossible to step without treading on the huge dark red flowers reared on stems only a few inches tall.' In 2006, Flanagan and Kirkham were able to report that, 'the floral carpet was every bit as striking as Wilson described it.'

Whether it is the Andes mountains, China, Taiwan or other 'hotspots', we must never assume that orchids are 'safeguarded in reserves' unless there are resources and incentives in place to protect what little remains. In a warming world, there will be more people desperate to eat cooked food, drink clean water and escape poverty. Conservationists should be prepared to enlist more local communities to be advocates of orchid reserves like the Mihary Soa in Madagascar, and to work with people in need. Otherwise, it may not be practical to safeguard what remains of the world's hotspots. But there are many kinds of threats to orchids. We must consider each before we proceed.

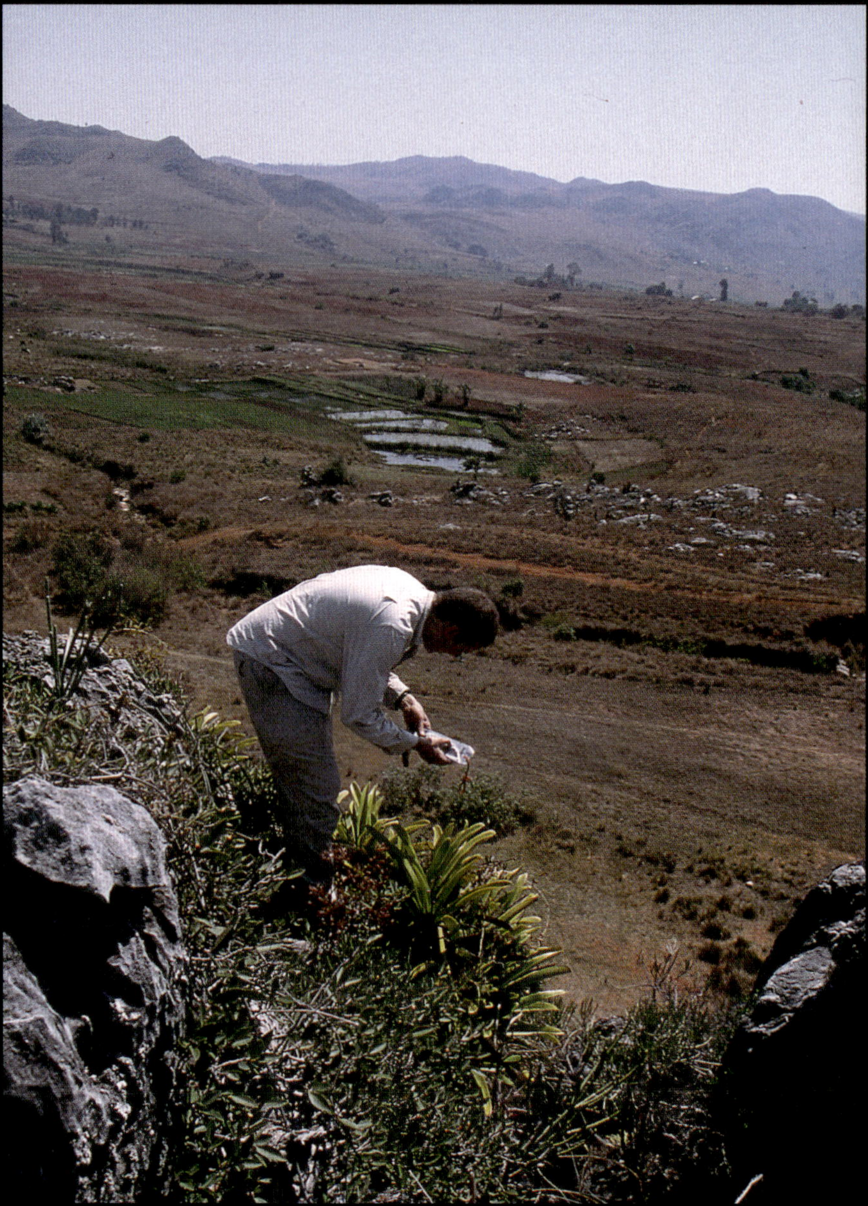

10 Orchids on the edge

'In Peninsular Malaysia habitat loss due to human activity has led to a serious reduction in orchid diversity in the region . . . as millions of hectares of virgin forest are annually clear-cut for agriculture, mining and urban development.' Rusea Go and Edward Entalai Besi, 2020

In 2020, Peter Raven and Scott Miller concluded that over the past 25 years, around a quarter of all tropical forests had been lost globally. The rate of that conversion is both astonishing and shocking. In 1991, Calaway Dodson and Alwyn Gentry described the situation in western Ecuador as 'grim'. They said that at the end of the 19th century more than 75% of the region was covered by primary (original) forest, but just 4.4% of that original forest remained. Moist and wet forest had declined from 50% of the region to less than 1%, and from 15% to 0.1%, respectively. Viewed from the air, the mountains of much of Central and South America appear as a patchwork of fields interspersed with the remnants of the forests that originally cloaked the land. At ground level the fields resolve themselves into pastures, sometimes with a few isolated trees remaining, often along the margins of streams. If only these living relics could talk: what stories would they tell us?

   Throughout the world, about half of all orchid species are threatened by the 'fearsome five': habitat destruction or land conversion; invasive species; loss of pollinators; poaching; and climate change. The threats vary from continent to continent, and from country to country. In Indonesia and other parts of tropical Asia, the conversion of forest into vast palm oil plantations is the main threat. In South America, it is the conversion of land for cattle and arable farming that are the primary threats. Tropical ecologist Diana Lieberman, who studied Costa Rican forests, once remarked, 'Just remember that a kitchen-sized chunk of rainforest was cleared for you to eat that beef hamburger.' These factors can interact with one another, with disastrous consequences for plant populations. In 2007, for instance, 18 orchid species fell victim to catastrophic fires in the elfin forest-mountain rainforest of Montebello in Mexico, and are now extinct. The fires were the result of habitat destruction for agriculture, followed by vegetation damage caused by severe frosts, and extreme drought leading to drying of the epiphytic flora. The role that climate change had in exacerbating these factors, which led to canopy fires previously unknown in such a wet area, is difficult to determine. What we do know, however, is that extreme climatic events are already becoming commonplace.

   Other threats to orchid habitats include illegal mining of precious metals, primarily gold, zinc, and copper. Gold mining, in particular, has a devastating effect on plant and animal habitats. It leaves a legacy of sinkholes, contaminated soil, and water polluted with choking sediment, mercury and other toxic chemicals that are carried downstream into villages and pristine areas. Damming of rivers leads to flooding of precious orchid habitat. The ongoing

(previous left hand page)
David Roberts monitoring a small population of *Angraecum longicalcar* **on a marble inselberg in Madagascar.**
Photo: Philip Seaton

(previous right hand page)
Angraecum sororium **growing as a lithophyte in Madagascar.**
Photo: Lawrence Zettler

destruction of the Amazon, mostly for cattle and soya, tends to monopolise the headlines, but it is also widespread in much of the rest of South America. In 1999, Juan Felipe Posada commented that, 'Destruction of native habitats for flora and fauna in Colombia is being done at such a rate that in very few years not much of the original forests will exist. On one side, the natural expansion of the population requiring more land for housing, agriculture or cattle projects, keeps tearing down original, untouched woods and forests. On the other side, and at an even higher rate, Colombia's drug problem is causing great devastation.' In Australia, we might add fire mismanagement to the list. Wherever and whenever human interests collide with Mother Nature, biodiversity loses out. Although extreme weather events capture the headlines, the gradual and relentless changes that are taking place are, in many respects, even more disturbing.

The living dead

In an age when ships were transported over oceans by wind alone, sailors noted how forested islands were cloaked in clouds, whereas those that lacked forests were not. The selective removal of valuable timber from forests can lead to sudden changes in the microclimate, contributing to the decline of sensitive species. Ironically, the diversity of tropical forests has contributed to their destruction. Why? Because the hardwoods most valued for lumber (for example rosewood and mahogany) are sparse in the landscape. Trees deemed unprofitable for their lumber are felled and dragged away by cable, loaded onto trucks, and shipped off to powerplants or to make charcoal. The few small trees that remain in the hot, sun-baked landscape poke upward towards the sky; most are bent or tilted at an angle as a reminder of the destruction.

Travelling along the road south of Neiva in Colombia, occasional lone forest giants full of *Cattleya trianae* remain as remnants of the original forest. When the trees eventually die, they will take with them the myriad of epiphytes they support on their branches, never to be replaced. Writing in *The Orchid Review* in 1992, David Miller and Richard Warren recalled that they had driven the road from Papucaia to Cachoeiras de Macacu in Brazil over a thousand times during the previous 25 years. 'It is a 15-mile-long section where you leave the Rio de Janeiro coastal plain and enter the foothill country of the pre-sierra. In the early days there were still original forest flatlands

> As you drink your early morning cup of coffee, you might take a moment to reflect where it was grown. The ideal altitude for coffee cultivation is between 900 and 2,000 m above sea level (2,950 to 6,550 ft). Coincidentally, this same altitude lies in the range at which you find the greatest diversity, not only of orchids but also of many other unique plants and animals. Should we reduce our intake of coffee to promote orchid conservation? Perhaps. But as coffee is a shade-loving species, some plantations are situated beneath large trees that harbour orchids. Coffee is also bee-pollinated, and at least one study revealed that coffee yields are higher in plantations that border intact forests that provide a home for these bees.

either side of the road; remnants, but sturdy remnants, some being around five or six hectares. We well remember during early December children selling, for a few cents at the roadside, plants of *Cattleya harrisoniae*, a beautiful rose pink species, with large flowers and *Oncidium flexuosum*, a small manyflowered yellow gem. They were plucked from the forest remnants where they grew in profusion. During the past eight years no trace of either have we seen, as there are now no original remnants.'

'Except for this year. In a five hectare roadside pasture, which, either by mistake or by some whimsy of providence, still has standing a dozen or so large original trees dotting the field, we saw a huge clump of the *Cattleya*. Over four hundred flowers on the topmost branches of a large dying tree. A glorious mass of light purple and pink; a classic colony; as rich a sight as we'd ever seen. And there they were. The last of their kind. DOOMED! No specific pollinators left. No symbiotic fungi around. No 'friendly' tree for miles. No unchanging climax forest. No stable environment. No hope. No way. DOOMED! But they were very beautiful, quite worth a tear and quite worth the object lesson they demonstrated of what mankind loses on a daily basis from ignorance, greed and thoughtlessness. Doomed!'

(above)
Remnant of forest growing by a stream in Ecuador.
Photo: Philip Seaton

(opposite)
Endemic to Colombia, *Cattleya quadricolor* is one of the most endangered orchids in the Cauca Valley.
Photo: Philip Seaton

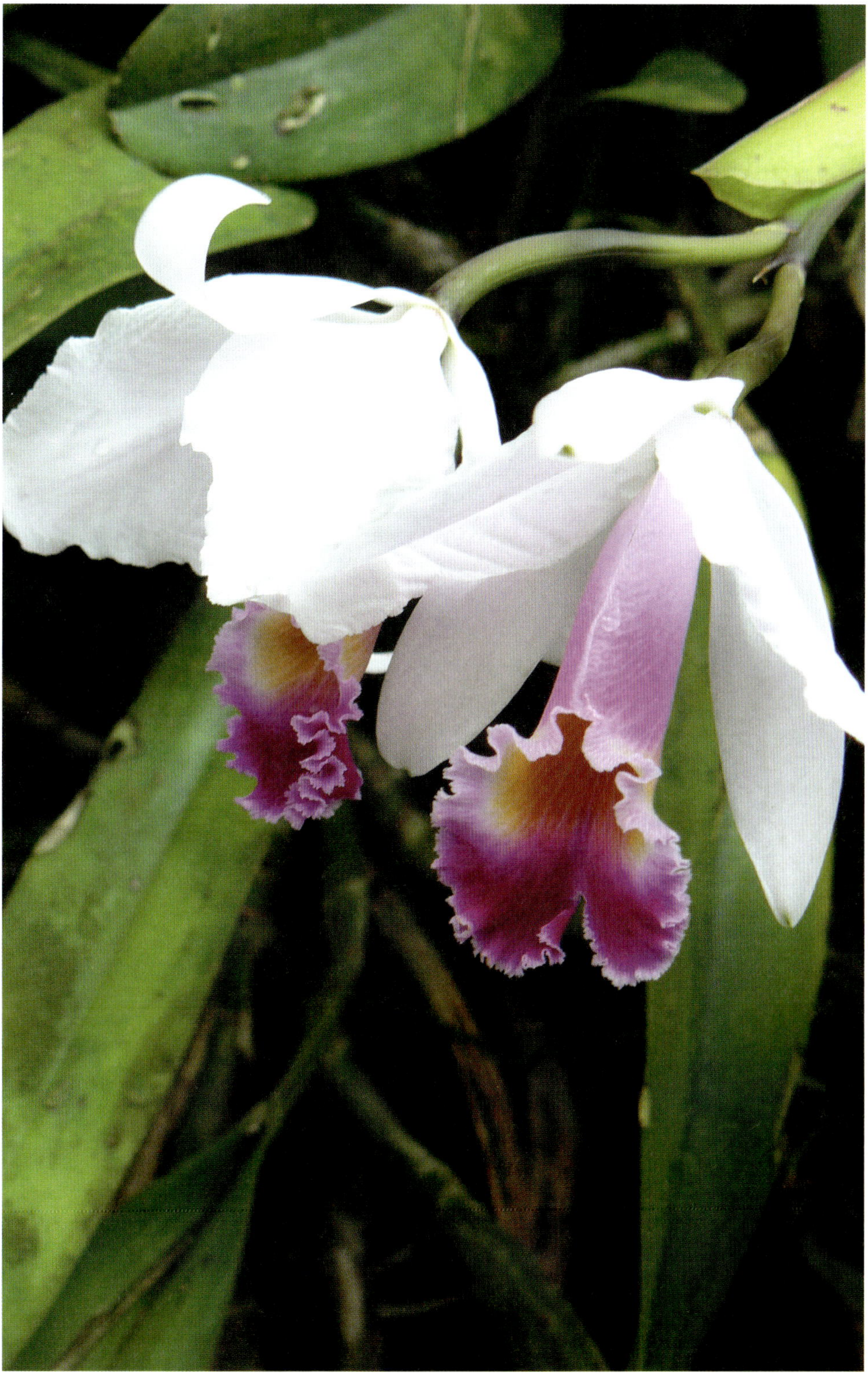

Road and trail building
'We rode over the recently opened link of the Pan-American Highway from San Isidro to Cartago, [Costa Rica], making the journey in a little over three hours – the same journey ten years ago, before there was a highway, I made in five days, afoot. But it was a more profitable and memorable experience to go over El Cerro de La Muerte on foot, travelling slowly, and stopping to examine whatever plant, bird or vista claimed attention.' Alexander Skutch, 1946

Ironically, the construction of new roads (legal and illegal) provides access to remote areas, leading to the discovery of new species, while at the same time allowing the same species to be collected illegally. In 2000, Phil photographed a large clump of *Masdevallia rosea* growing on a roadside tree between Quito and Papallacta in Ecuador. Returning a few years later, the tree was still standing, but the masdevallia was long gone. Visitors to Viñales in Cuba who spot the pretty *Broughtonia lindenii* growing high up in some ancient roadside trees may conclude that this is their preferred habitat. The truth is that growing high up in the canopy, they are out of reach of would-be collectors. The orchids that grew further down the trunk have been 'harvested'.

Natural disasters
In addition to the destruction caused by humans, orchid habitats are vulnerable to natural disasters. This is especially evident in the tropics, where cyclone activity is prevalent. As predicted by climatologists, the severity of cyclone activity is increasing in our warming world. In western Cuba, and throughout the Caribbean, the effects of increased cyclone activity have already been documented. In 2004, Hurricane Ivan – a Category 5 'monster storm' – passed directly over Guanahacabibes National Park, resulting in the loss of 60% of all ghost orchids rooted on trees. Researchers in Spain and Cuba concluded that *Dendrophylax lindenii* could succumb to extinction on the peninsula in a mere 25 years if the annual probability of disturbances exceeds 14%. The same would be true on the nearby island of Grand Cayman, where another kind of ghost orchid (*D. fawcettii*) is endemic. These leafless ghosts are Caribbean neighbours, with *D. lindenii* found in Western Cuba and *D. fawcettii* around 350 km (217 mi) to the southwest in the Cayman Islands. Both occupy low-lying coastal habitats, making them prone to catastrophic damage by annual tropical cyclones. Grand Cayman's highest point, for example, is just 18 m (59 ft) above sea level, and the same is true for Guanahacabibes, Cuba.

Unlike the ghost orchids in Cuba and Florida, the labellum of *Dendrophylax fawcettii* lacks the distinctive long-branched twin 'tails' that dangle downward and appear to dance in the slightest breeze. In 2012, it was featured in a book by the IUCN and

Another ghost orchid, *Dendrophylax fawcettii*, is endemic to the Cayman Islands.
Photo: Stuart Mailer

Zoological Society of London called *Priceless or Worthless* as one of the world's one hundred most endangered plants and animals. It is confined to just 2.4 hectares (6 acres) in a small (18.6-hectare/46-acre) fragment of old growth ironwood (*Chionanthus caymanensis*) forest, surrounded by urban development. The area was given a stay of execution in 2008 when plans for a bypass through the area were abandoned, due to a campaign by local residents who were opposed to the new road.

Inland areas are not immune to tropical cyclones. In 2008, four years after Ivan, Hurricane Gustav lashed Cuba with torrential rain and winds of 240 kph (150 mph), devastating Jardín Botánico Orquideario Soroa, which houses the island's most important collection of native orchids. Most of the orchids growing in the shade houses were damaged, and around half were lost. No area of the garden escaped the wrath of the storm. Windows and doors were broken, destroying equipment, leading to the loss of many important documents. The roof of the main house was shattered by falling branches and trees. The laboratory, which housed an embryonic seed bank – a critical component of the drive to conserve the Cuban orchid flora – was inundated with water. Luckily, no flasks were broken, but subsequently there have been many problems with contamination by microbes. Today, the garden has largely been restored, but remains vulnerable to cyclones and electrical blackouts from fuel shortages.

Then there are orchids that grow on the slopes of active volcanoes. In 2005, large tracts of pristine cloud forest were destroyed by repeated eruption of the Soufrière Hills volcano on Montserrat in the Caribbean. The island harbours 800 native plant species, including the endemic and aptly named *Epidendrum montserratense*. A year after the eruption, Kew conservationists rescued several living individuals of this endangered orchid from dead mango trees, and placed them in the security of Montserrat's newly developed botanic garden, where they are being cultivated for public display. Seeds have also been safeguarded in cold storage at Kew's Millennium Seed Bank in the UK. Montserrat serves as a reminder that some level of conservation is possible, even after a volcanic eruption.

The urban environment
All too often, the urban environment is overlooked as a new home for orchids, and the opportunity those plants can provide for local people. With its mosaic of suitable habitats, the Italian municipality of Cagliari, for example, contains one quarter of Sardinia's 62 orchid species, including a subspecies, *Ophrys exaltata* subsp. *morisii*, that is endemic to the island. However, climate change, combined with the urban heat island effect (where towns and cities are hotter than the surrounding countryside), can lead to droughts, putting individual species at risk of future local extinctions.

Not only can orchids be found in local nature reserves and parks, many species often appear unexpectedly on small pockets of uncultivated land and in gardens in the UK. High Wycombe, a small town a few miles north of London, is within the Chiltern National Landscape – a chalk escarpment that boasts a particularly rich orchid flora. This includes common spotted (*Dactylorhiza fuchsii*), pyramidal (*Anacamptis pyramidalis*), bee (*Ophrys apifera*), fly (*Ophrys insectifera*), and green-winged (*Anacamptis morio*) orchids, which grow in a number of the small reserves scattered through the town and its outer fringes. Chiltern Rangers, a not-for-profit organisation, have instigated a number of projects in the town that take advantage of the fact that a simple change in land management, such as cutting the grass and removing the mowings to reduce competition, can quickly result in increased orchid populations. A donation of 96 green-winged orchids from Kew is being maintained at the project's Wildflower Nursery – the staff having received training from Kew on their upkeep. The aim is to introduce the orchids into the grounds of local schools. One school already has wild green-winged orchids growing in its grounds. Public engagement and community participation lie at the heart of the project, and involving schoolchildren provides additional educational opportunities. The longer-term aim is to plant orchids

India's 'Golden Paph Orchid', *Paphiopedilum druryi*. Grown by Mark Turner.
Photo: Lawrence Zettler

Iris Suarez and *Dichromanthus aurantiacus* in the Reserva Pedregal de San Ángel, Mexico City.
Photo: Philip Seaton

in a wide range of different types of urban community locations, such as hospitals, sheltered housing accommodation and parks, to create wildlife gardens where students of all ages, abilities and backgrounds can get involved.

The botanist Viswambharan Sarasan is developing a five-year urban project on behalf of Kew in the state of Kerala in India. Although its land mass comprises just 1% of India as a whole, Kerala is home to around 30% of that country's orchids, and contains many endemic species, including the critically endangered 'golden paph' *Paphiopedilum druryi*. Kew is joining hands with the Kerala State Council for Science, Technology and Environment, Kerala Agriculture University, the Centre for Biodiversity Conservation, and seven colleges. The aim of the programme is to rescue orchids from trees that are destined to be felled as part of infrastructure projects. A total of 25 different orchid species have been targeted for collection. These will be planted in urban areas where, taking advantage of their charismatic nature, people of all backgrounds will have an opportunity to engage with their native orchids. Species that are rare but not available in large numbers will be propagated using capsules collected from diverse sources, and planted on to urban trees and college campuses. Students will plant the seedlings and care for the plants. They will also be given opportunities to carry out projects on different aspects of the *in vitro* conservation of orchids, including microbiome assessments of organic matter on host trees, which will allow them to understand the community composition and abundance of key mycorrhizal groups. A school programme is being developed with the help of KSCSTE that – depending on the availability of funds – could be rolled out in many schools in the coming years. The goal is to develop several conservation, engagement, and education projects in the state with support from the state and central governments.

In Mexico City, one of the world's largest cities with nearly 22 million residents, native orchids have adapted to the urban environment and can be encountered in some surprising places. *Pedregales* are old lava fields, where the cracks and crevices in stony or rocky ground provide a home to a suite of plants that you would not normally find growing together. The Reserva Ecológica del Pedregal de San Ángel contains the remnants of vegetation that were present 1,800 years ago after the eruption of the Xitle volcano. Located on the campus of the Universidad Nacional Autónoma de México (UNAM), it is home to 24 species of xerophytic (drought-adapted) orchids that grow alongside scrubby vegetation, including cacti (*Opuntia streptacantha*) and dahlias. Almost all the orchids are terrestrial. Recent urban expansion has led to their original habitat being almost completely covered by asphalt, and they are under pressure from air pollution, acid rain, frequent fires, and increasing temperatures. Scientists at UNAM are not only aware of this threat, they are actively working to conserve their orchids.

Pilar Ortega-Larrocea and her students have been studying the orchids in the reserve for around 25 years. In 2000, they initiated a reintroduction programme, beginning with the endemic and endangered terrestrial, *Bletia urbana*, which included the use of several different mycorrhizal fungus isolates acquired from the orchid's natural habitat in the Mexican countryside. Soon after collecting seeds, they used these fungi to grow seedlings in the laboratory that were subsequently outplanted. Some of the *B. urbana* seeds from their initial experiments were maintained in cold storage for later use. After 14 years, the embryos of these seeds retained their viability and are still used today for reintroduction efforts. Based on annual surveys, the survival rate of the outplanted seedlings is around 50%.

Seedlings of the hummingbird-pollinated *Dichromanthus aurantiacus* have also been reintroduced. The success of widespread species such as *D. aurantiacus* and *D. coccinea* (widespread throughout Mexico) may be due to their association with fungi (*Ceratobasidium*) that also grow quickly and easily *in vitro*, and probably also in soil. By contrast, *Bletia urbana* is associated with different fungi (*Tulasnella*) that are slower growing and probably more restricted in their soil niche requirements. Suitable microhabitat conditions have also been shown to be crucial for seedling survival. Seedlings raised with fungi have higher survival rates (75%) than those grown without (12%) in the first year after reintroduction.

Most recently, Ortega-Larrocea's team has been investigating *in situ* germination of their native lady's-slipper including the beautiful golden *Cypripedium irapeanum*, reporting with excitement, 'It is amazing how *Cypripedium* takes 10 years to germinate *in situ*! I cannot believe after 10 years of baiting we still have the baits *in situ* and found again!' Data analysis of herbarium specimens indicates that populations of *Cypripedium* species in Mexico were previously much more widespread. Though long since dead, herbarium specimens, and the information written on their labels, serve as windows into the past, giving us insights into the habitat requirements of various orchids species. Collectively, Ortega-Larrocea's experiments with *Bletia* have led to a better understanding of the factors that promote orchid restoration, and the knowledge can now be applied to other orchid habitats in urban areas.

Illegal collection

The urge to collect seems to be embedded in the human psyche: stamps, coins, fossils, books, teddy bears, orchids, the list of collectables is endless. For the most part, collecting is a harmless pastime, but for a few individuals, it can become an obsession: the desire to have becomes the desire to have at any price. For orchid obsessives, as soon as a new species is discovered, the 'must-have' desire kicks in. For some people, the rarer a species, the more desirable it becomes, setting up a vicious cycle of increasing rarity adding pressure on the species' existence in the wild. Why some people cannot wait for seed-raised material to become available defies both logic and ethics.

Guatemala exemplifies how illegal collection has contributed to orchid decline. This Central American country is perhaps best known in the orchid community as the home of *Lycaste skinneri*. Huge quantities of *L. skinneri* were imported into Europe by George Ure Skinner and his successors. Six shipments were included in Loddiges' catalogues from 1841 to 1843, and in 1846, Lindley wrote about large masses of *L. skinneri* for sale. Reports of consignments of 100,000 plants sold at auction towards the end of the 19th century attest to the ransacking of habitats simply for the sake of commercial greed.

What is perhaps less well-known is that Guatemala is a megadiverse country. With around 1,200 species, it has a rich and diverse orchid flora that includes 211 pleurothallid species. Because they lack pseudobulbs to tide them over during periods of drought, pleurothallids are vulnerable to dehydration. This, together with their habitat and host specificity, makes them particularly valuable as indicators of declining environmental health, including as a result of climate change. Typically found in moist forests between 1,800 and 2,800 m (5,900 and 9,200 ft), pleurothallids show a high degree of endemism. Sadly, Guatemala lost half of its forest cover between 1950 and 2002. Their restricted distribution, combined with the usual suspects of anthropogenic pressure and illegal collection, means that many pleurothallids are at risk of exinction in their native habitats. In addition to belonging to a niche market for plant collectors and tourists, local people purchase orchids for special occasions, such as birthdays and weddings. Commonly thought to be parasitic (epiphytes are often referred to as *parasitos*), they are scraped off the branches of small trees in the mistaken belief that they harm the trees – a lingering and misleading perception globally. Principal among the endangered species are *Lepanthes*: 67 of the 70 Guatemalan species of these tiny jewels of the cloud forest are the most frequently sold pleurothallids. There is a growing market due to increasing numbers of collectors, who extract plants from the forest, bringing many of the showiest species closer to the edge of extinction.

The illegal collection of orchids is not confined to the tropics. The terrestrial orchids found in temperate regions can be equally vulnerable, as they are easier to spot and remove than their epiphytic cousins in the tropics. Although we do our best to encourage exploration of the natural world, and want people to enjoy their encounters with orchids, there is the ever-present danger that someone will dig up a plant, especially if it is rare. The experience of one of the authors (Phil) in the UK is by no means atypical. Walking through his local woodlands, much to his surprise and delight, he stumbled upon a long raceme of white flowers . . . a sword-leaved helleborine (*Cephalanthera longifolia*) – growing by the side of the path. Returning the following day with a camera, he found that the plant had been dug up. Not only does such a selfish activity deprive other people of the joy of finding such a rare (in the UK) woodland gem, but the chances of the plant surviving for any length of time in cultivation were zero, barring a miracle. Such unethical and widespread practice forces naturalists to keep the locations of many orchids a closely guarded secret. Can you blame them?

In Switzerland, a healthy population of more than 2,000 plants of *Cypripedium calceolus* was reduced to three individuals by poachers. In Romania, a single *C. calceolus* var. *citrina* specimen was found among a large population of the 'normal' form, only to be dug up. Illegal collection for medicinal purposes may be the greatest threat in many countries. Stephan Gale recently reported in *The Orchid Review* that more than 400 species of illegally collected orchids were being traded at markets in southern China, totalling more than 1.2 million plants worth at least $14.6 million on the black market. In today's technological age, orchids may be sold illegally via websites or social media. A new species, *Paphiopedilum*

Evidence of illegal orchid collection in the Cantareira State Park, São Paulo, Brazil.
Photo: Luciano Zandoná

nataschae, for example, was discovered in May 2015, and a flowering specimen was almost immediately offered for sale on the internet in November the following year.

In all probability, the local collector receives very little financial reward, whereas it is the unscrupulous middleman that makes the most profit. The commercial extraction of plants from their natural habitats not only endangers wild populations, but makes it nearly impossible for plants that are legally propagated in nurseries to compete with the lower prices of wild-collected plants. It can take many years for an artificially propagated orchid to produce its first flower, and a considerable investment from the grower in time, expertise and money may be required to produce adult specimens for sale.

You might have thought that plants would find refuge in botanic gardens, but this is by no means always the case. We were shocked to learn that 900 newly transplanted orchids had been stolen early in 2023 from a conservation garden at Kings Park and Botanic Garden in Western Australia. The thieves took 400 critically endangered Carbunup king spider orchids (*Caladenia procera*) and 500 Collie spider orchids (*Caladenia leucochila*). It

Illegally collected orchids for sale outside an exhibition in São Paulo, Brazil.
Photo: Luciano Zandoná

is heartbreaking for the conservationists who have invested so much time and effort in producing plants for reintroduction, and who want to share their passion and the fruits of their labours with the general public. At the Naples Botanical Garden in Florida, where artificially propagated ghost orchids are on display on trees along a boardwalk, security cameras appear to have discouraged unscrupulous members of the general public from removing the orchids illegally.

 What, then, is the answer? Hobbyists and growers should routinely question the origin of the plants offered for sale. They should only buy from reputable sellers. Particular caution should be exercised when buying online. There are people who believe there is nothing wrong with buying cheap orchids via the internet without the need for CITES import permits and phytosanitary certificates (see 'International trade', page 149). Purchasers should ask to see the evidence that the plants were obtained legally. Otherwise, every plant bought on the internet without the accompanying documentation is undermining the legitimate orchid trade that follows the regulations. Every dollar or pound spent could be regarded as one vote rewarding bad behaviour. Was the plant nursery grown or collected from the wild? Is it a division of a plant that has been in cultivation for many years? Were the plants grown from seed? Illegally acquired plants do not produce legal seed. Flasks of seedlings and seedlings in pots must have been raised from seed, and laboratory-raised plants are more likely

Laelia speciosa has been sold in markets in Mexico for many generations. This painting is based on a historical photograph, probably taken some time in the 1940s. Painting by Philip Seaton

to survive in cultivation, as wild-collected plants are frequently difficult to establish.

Illegal collectors beware! Molecular techniques can now reveal the identity of an orchid collected from the wild and trace its specific origin. Thus, poachers who sneak into natural habitats and illegally sell orchids on the black market may have a lot of explaining to do to law enforcement officers, as will buyers. The culprits will also have two potential mug shots to look forward to, one taken by motion-activated digital cameras secretly placed in remote habitats that document the offence and offender, and the other taken before they serve their jail sentence.

Laelia speciosa on the brink

With large rose-lilac blooms, *Laelia speciosa* is one of the most beautiful and charismatic Mexican species. Since pre-Columbian times, it has been regarded as one of Mexico's most important orchids, from both a cultural and a horticultural standpoint. Despite being widely distributed and with some large populations remaining, its long-term survival in the wild is at risk. This is largely due to the destruction of the oak forests where it grows epiphytically, but the illegal removal of plants for either commercial or religious purposes also contributes to its decline.

Between May and June, plants are offered for sale in markets in many Mexican towns. Trucks arrive at local markets with boxes of flowering shoots and plants, and the orchids are passed on to street vendors. It has been estimated that during the flowering season, between 10,000 and 100,000 plants or flowering shoots arrive in Mexico City, which is the main centre for the distribution of *Laelia speciosa* and many other orchids. In the state of Veracruz, a region where the species is not known to occur naturally, around 10,000 are sold annually. In Michoacán, around 6,000 plants are sold each year, as are an unknown number of flowers that are taken from the wild populations for religious festivals. Unless the demand for wild material drops, there is little incentive for *L. speciosa* to be artificially cultivated. In places where new flowering shoots are removed from a population, this has the effect of reducing the plants' vigour, contributing to their physical decline. Consequently, the orchids are unable to produce flowers during successive years, reducing their fecundity. They gradually become scarcer, increasing the risk of local extinction.

International trade

In the early 1970s, wild-collected orchids were still commonly advertised for sale in journals, and could be purchased from many orchid nurseries in Europe, the USA and Japan. Everything changed with the establishment of CITES (the Convention on International Trade in Endangered Species). CITES regulates the international trade of plant and animal specimens from the wild,

Today, *Laelia speciosa* is still sold illegally, and is increasingly at risk of extinction in its natural habitat.
Photo: Mariana Hernández

and remains controversial in many quarters, where it is often seen as being unnecessarily bureaucratic and a minefield for the amateur grower. It can make the legitimate exchange of scientific specimens difficult, and arguably in some circumstances, it is an obstacle to conservation. Nevertheless, its aim of preventing the illegal trade in orchids is to be applauded, and all responsible traders and growers must observe the regulations.

As they are a globally imperilled family, all orchid species are listed under either Appendix I or Appendix II of CITES (http://www.cites.org). Appendix I lists species that are thought to be threatened with extinction, and includes all genera of slipper orchids. Trade is subject to strict regulation and is only authorised in exceptional circumstances. Appendix II species are not currently threatened with extinction, but are expected to become so unless trade is strictly regulated. The raising of orchids from seed is encouraged, provided that the seed has been obtained legally and that its harvest does not have a detrimental effect on wild populations. The hope is that this will reduce the pressure on wild populations as a source of material. Seeds, capsules, and pollinia of Appendix II species are exempt from CITES, whereas those in Appendix I are not. Plant materials should be accompanied by a phytosanitary certificate from the country of origin. In the UK, this should be affixed to the outside of the package with 'FAO HM Customs and Excise' visible, along with a letter of explanation about the contents (species, provenance, and how cultivated). The general protocol also applies in the USA.

The Convention on Biological Diversity (CBD) establishes the principle that nation states have sovereign rights over access to their own genetic resources. Putting it another way, you wouldn't expect to be able to go to another country, harvest orchid seed from the wild, and bring it home any more than you would expect to be able to walk into someone's greenhouse and harvest orchid seed without their permission. If contacted by someone in another country who says, 'I can send you orchid seed without any problems from my cultivated plants.' The correct response is, 'What are the rules at your end for exporting of orchid seed, pollen or seed capsules?' The CBD encourages seed banking because it requires states to adopt measures that promote the establishment and maintenance of facilities for *ex situ* conservation.

Flasks of orchid seedlings or tissue cultures obtained *in vitro*, those in solid or liquid media, and material transported in sterile containers are exempt (for species listed by CITES in either Appendix I or Appendix II). The UK and many other countries require a phytosanitary certificate from the exporting country. Orchid plants not in flasks can be traded legally so long as the regulations are observed. In the UK, a CITES export permit is required from the country of origin. This is then used to obtain an import permit from the Department for Environment, Food & Rural Affairs. In the USA, all import permits (phytosanitary certificates) are obtainable only after prior acquisition of a CITES permit. Phytosanitary certificates are often required by both the exporting country and the importing country to avoid both the export and import of pests and diseases. This certificate travels with the shipment, and copies are also sent in advance to authorities in the importing country, such as the UK's Animal and Plant Health Agency, noting the time and place of arrival. Before samples are allowed to leave the importing country's airport, a physical inspection must take place in addition to other checks that sometimes require a fee (for example, customs inspections and taxes).

To illustrate the correct procedure, a ghost orchid grown under the care of Johanna Hutchens at the Chicago Botanic Garden was chosen as the 'crown jewel' for an orchid conservation exhibit at the Chelsea Flower Show in London in May 2023. But before the orchid could be physically transported across the Atlantic Ocean, CITES permits were required both in the USA and the UK, and these had to be in hand during the journey. Authorities at Heathrow Airport were informed of the orchid's arrival ahead of the flight. When it arrived at the airport, together with a phytosanitary certificate issued by the US authorities, it was subject to inspection by the Animal and Plant Health Agency.

Climate change

'If nothing is done about climate change, you can forget about biodiversity.' E. O. Wilson, 2019

In the centre of the small market town of Spilsby in the UK, on the edge of the Lincolnshire Wolds and the fenlands, stands a statue erected in the memory of Sir John Franklin. The two ships (HMS Erebus and Terror) and crew under his command were lost in 1847 while exploring the Canadian Arctic. They were seeking a route around the top of North America, the 'North-West Passage', only to become icebound for over a year. Lost, lonely and cold, with dwindling supplies, most of the crew along with Franklin died horribly. The survivors abandoned their ships and walked over the ice towards the Canadian mainland, never to be seen again. Had the expedition occurred 150 years later in our warming world, their ships would have sailed onward with little resistance from floating ice. At the current rate of warming, a totally ice-free North Pole during the summer months is now an alarming and real possibility.

As we write, the year 2023 has been confirmed to be the warmest on record. Our planet is expected to reach or exceed 1.5°C (2.7°F) of warming above pre-industrial levels within the next 20 years, regardless of how profoundly we cut greenhouse gas emissions. Parts of the world are seeing an increase in the frequency and severity of heat waves and droughts. A problem in themselves, prolonged droughts make wildfires more likely. Droughts and a gradual drying of the climate in Australia have undoubtedly contributed to the horrific recent fires there. As we have seen, orchids have been lost to fires on the other side of the globe in Mexico, and presumably to the north in California. How will climate change affect orchid populations over the long term? How might they respond to climate change? It's not just that temperatures are ticking upwards, it is also the rapidity with which the warming is taking place that is problematic. Human society may be able to adapt, but plants in general may not, especially those that, like orchids, have complex life cycles that are closely tied to other life forms.

Tropical forests in general face an uncertain future. Cloud forests are uniquely threatened both by human pressures and by climate change, because of the projected impact the latter will have on temperature, rainfall, and cloud formation in mountainous regions. Species adapted to life on mountainsides at lower elevations face the prospect of losing their habitat, and must retreat to higher elevations if they are to survive. One such area of interest lies in biodiverse Costa Rica. Over the past two decades, Fern Perkins has seen many changes in the Monteverde Cloud Forest,

1,400 m (4,600 ft) above sea level. The most obvious change has been the lack of persistent cloud cover spilling over the mountains from the Atlantic side, bathing their peaks in mist. As a result, there is less moisture cloaking the cloud forest and more sunlight trickling in from the west, heating up the air in forests around the region. Reptiles respond to the increased temperature by migrating to higher elevations that were once off limits to the cold-blooded creatures. The deadliest venomous snake of the Western Hemisphere, the fer-de-lance, *Bothrops asper*, rarely slithered higher than 1,000 m (3,300 ft) above sea level because it was too cool for it to live at those altitudes. But it is now seen regularly in Monteverde at 1,440 m (4,660 ft), and in 2022, one of the town's residents was fatally bitten.

In northern temperate climates, rather than migrating to higher elevations, the question becomes whether orchid populations will be able to migrate north quickly enough. Inevitably there will be winners and losers. In the UK, the southern marsh orchid (*Dactylorhiza praetermissa*) and the bee orchid (*Ophrys apifera*) have both increased in frequency, and their ranges have pushed northward over the past 20 years. Over the same time period, the ranges of the burnt orchid (*Neotinea ustulata*) and the lesser butterfly orchid (*Platanthera bifolia*) have contracted. The reasons for these changes are likely to be complex, but the warming of the British climate undoubtedly has played a role. In the USA, the floras of northern states may become 'richer' as southern species move in, but those that move out will not be random. The diminutive North American terrestrial fairy slipper (*Calypso bulbosa*) is apparently one cold-loving species 'on the move' northward into Canada, as it is becoming noticeably more scarce in the continental USA year by year.

In addition to changes in range, there have been changes in flowering times. Britain and Ireland have a long tradition of amateur naturalists and recorders of wildlife, stretching back to the 18th century. Using records for 406 plant species dating back to 1753, Nature's Calendar, a citizen science project, has shown that warmer temperatures are triggering earlier flowering in many spring plants. Between 1952 and 2019, plants have been flowering an average of 5.4 days earlier per decade. Michael Hutchings carried out a population study from 1975 to 2006 which showed that, on grasslands in the south of England, the early spider orchid (*Ophrys sphegodes*) flowered roughly half a day earlier per year – 5 days earlier per decade – during this period. In 1853, Henry David Thoreau noted that the first flowering of the pink lady's-slipper (*Cypripedium acaule*) around his home in Concord, New Hampshire, USA, typically occurred in late May. He would have been astounded to find that it now flowers around three weeks earlier.

Orchids that rely on specific insect pollinators are potentially vulnerable to any mismatch between insect emergence and flowering time. As an example, the male solitary bees (*Andrena nigroaenea*) that fertilise the early spider orchid (*Ophrys sphegodes*) normally emerge slightly before the females. If environmental warming leads to earlier emergence of the females, understandably, the males may become more interested in pollinating the females, and not the orchids. Timing is everything, and climate change has the potential to disrupt this delicate balance. In southern England, about 10% of the flowers of the early spider orchid set seed each year. In populations throughout continental Europe, seed set is usually much higher, at around 50%. This difference in reproductive success suggests that the population in England is pollinator-limited, and may be indicative of such a mismatch. Orchids in tropical latitudes may also suffer from similar ecological mismatches. Future predictions of climate-change are based on mathematical models that are constantly being updated, and it remains anyone's guess how closely the projections will align with what actually occurs in the future. Although climate scientists predict that warming will be more pronounced in temperate regions compared with the tropics, orchids and their biotic agents, including mycorrhizal fungi and other soil microbes, will experience dramatic change everywhere.

Occasionally, species that are native to the Mediterranean crop up in counties along the south coast of Britain, and climate change may bring with it the enticing prospect of new additions to Britain's native flora. A sawfly orchid, *Ophrys tenthredinifera*, recently created a stir when it was found in Dorset in 2014. Similarly, a colony of the tongue orchid, *Serapias lingua*, was spotted in the county of Essex. In June 2021, *Serapias parviflora*, the small-flowered tongue orchid, was found growing in the City of London, in the 11th-floor rooftop garden of the Japanese investment bank Nomura. How did these species arrive? Did seed of the sawfly orchid blow across the English Channel? Did someone plant it? Did the two *Serapias* species arrive in soil? The field in which *S. lingua* was found had previously been a strawberry field, and the seed could have come across with strawberry plants from Spain.

Whether such exotic species will become established in Britain will depend on a suite of environmental factors, including whether or not suitable mycorrhizal fungi and pollinators are present. Even if new species do become established, several barriers exist to impede the migration of plants across the landscape, including ever-expanding towns and cities, and vast cereal monocultures. On the other hand, it is not beyond the realms of possibility – if certain tipping points are passed – that the Gulf Stream could cease to function, and thus the UK could become colder.

Sea level rise
As ice sheets and glaciers melt and the water in the warmer oceans

expands, sea levels are rising at an alarming rate. According to NASA, sea level is rising by 34 cm (13 in) per decade – 340 cm (130 in) in a century – and the rate is accelerating. At the time of writing, the 'doomsday glacier' in Antarctica is projected to slip into the ocean before too long, adding a noticeable rise. What will become of the orchids that occupy coastal regions, such as the 'worm vine' (*Vanilla barbellata*) that clings to branches of red mangrove along Florida's south western coasts? Will they migrate, either naturally via seed or artificially through human intervention, or will they slip into the sea of extinction?

Many low-lying, orchid-rich areas are vulnerable to sea level rise. The ghost orchids of Florida and Cuba could be the first to succumb to coastal flooding. According to NASA, southwest Florida will experience substantial flooding by the end of this century, meaning that habitat preservation will not be a viable option for safeguarding the orchids in peril. Can a water-impenetrable wall be constructed along Florida's coastlines to restrict flooding in a similar manner to those in the Netherlands on the other side of the Atlantic? The answer is probably 'no' for the simple reason that Florida's bedrock is primarily porous limestone. Essentially, rising water on one side will seep into the other side from below – limestone acts like a sponge, channelling water through capillary action. Florida as we now know it is in trouble, and novel solutions are needed to prevent flooding of the Sunshine State's low-lying areas. Other low-lying areas include the Amazon region of South America, coastal Bangladesh in Asia, and islands throughout the western Pacific such as Palau. Orchids and their biotic associates will need to migrate to higher ground, and many will need our help in a process that has become known as 'assisted migration'. There are the 'purists' who continue to believe that all natural habitats should be left alone from all forms of human intervention, but aren't humans responsible for climate change? So how do we conserve ghost orchids in the most flood-prone region of Florida? Getting people to take notice is a start.

Drought, fire, and floods

As the world warms, the atmosphere will be able to absorb and store more moisture. Record temperatures and droughts, followed by devastating floods, have become the new normal in many countries. In September 2022, the Australian Network for Plant Conservation selected 11 species found in north-eastern New South Wales, with the aim of documenting the impact of fires by establishing the geographic distribution and size of the remaining populations after the 2019–2020 Black Summer fires, which has become known as worst bushfire season on record. The fires followed several years of drought that had led to an accumulation of dry material, followed by higher than average temperatures that ignited devastating fires across much of the state. In parts of the western USA, recent fires have become so hot that the subterranean flora and fauna were essentially cooked, rendering the soil infertile.

The year 2023 saw an unprecedented series of weather events. Heat waves, floods, fires, and record high temperatures were recorded in North America, the Mediterranean, China, and throughout the world. It was Earth's warmest year on record. As we write, it has been forecast that 2024 may be even warmer. The hottest 10 years on record have all occurred since 2010. Prolonged high temperatures and drought dry out the soil to increasing depths, making it exceedingly difficult for plant roots to absorb water. Lack of water also shuts down vital ecological processes managed by microbes such as mycorrhizal fungi, including decomposition and nutrient transfer that orchids and other plants depend on. Although the underground tubers of 'winter green' orchids may survive, their associated fungi may not. Many kinds of higher fungi produce resistant spores that allow these organisms to survive adverse conditions, but it is not known if, and to what extent, orchid mycorrhizal fungi can withstand drought. Changes that occur below ground may not be initially obvious to us, but the disappearance of mycorrhizal fungi will have a negative impact on terrestrial orchid populations due to a lack of the fungal inoculum needed to generate spontaneous seedlings.

In China, Holger Perner reported that 2004 had been 'the last "normal" late spring/early summer season in Northwest Yunnan.' After 2005, every year had lacked most of the sporadic

(above)
The alarming effect of drought on the Cerro de La Muerte montane forest, Costa Rica, March 2024.
Photo: Diego Bogarín

(opposite top)
In March 2024, the mountains of Maltrata, in drought-stricken Veracruz state, were the epicentre of some of the worst fires seen in Mexico, threatening a population of what may be a new variety of *Cypripedium irapeanum*.
Photo: Orchidarc

(opposite bottom)
Remarkably, a small *Telipogon* species survived the severe drought in 2024 that affected the Cerro de la Muerte, Costa Rica.
Photo: Diego Bogarín

157 ORCHIDS ON THE EDGE

spring rains, and subsequently he had seen a much-reduced floral display in the region. In 2009, most cypripediums and many of the non-orchidaceous flowering plants showed severe drought stress. Either the plants did not emerge at all, or those that did emerge were stunted, and the flower buds had become trapped in the foliage while trying to open. By 2015, after another six years of drought, everything was bone dry, the ground crunching under visitors' boots. Most of the herbs were missing. Where once the grass had been a lush green, there was now a landscape of dry grasses in shades of brown or sickly yellow.

Perner's correspondents told a similar story. At another site, in the county of Shangri-la in Yunnan, the ground was described as being very dry. *Calanthes* were dying, and the humus on the forest floor was mixed with the dry silt. Again, the flower buds of cypripediums were seeming to 'freeze in place' as they were emerging, and most plants failed to emerge at all. Prior to 2005, eastern Tibet and western Sichuan experienced rain or snow almost every day. Botanists working in the field had to wear cold weather clothes most of the time they were outdoors during the months of July and August. Around 2005, the conditions began to change, with many days going by with bright sun and no precipitation. The fields of barley were drying up by early August. Botanists donned short-sleeved shirts almost every day. During their first ten years of field work, when they frequently travelled through Zhongdian on the way to Tibet or western Sichuan, the researchers would always stop to photograph and admire the extensive carpets of wildflowers all around the Zhongdian basin. As they had observed elsewhere, fields of wildflowers had dried up and disappeared, or had shrunk to just a fraction of what they once were. They write that today, no-one would believe just how extensive those colourful meadows had been.

Epiphytic and lithophytic species are particularly vulnerable to fire events because, unlike terrestrial orchids, they have neither an underground dormant phase nor a persistent seed bank that would allow populations to recover. For example, a recently discovered population of *Plectorrhiza purpurata* in Australia was heavily impacted by drought, with most host plants dead along with their associated epiphytes. In March, 2024, Diego Bogarín tells us that he spent some time in the montane forest of Cerro de La Muerte. A combination of El Niño and global warming meant that rainfall had been unusually low, resulting in warmer and drier conditions. For the first time in his life, he was confronted by burning and dying vegetation in what was normally a cool, moist habitat. Remarkably some telipogons survived, thriving on the leaves of a dead, dry *Blechnum* fern – a testament to the resilience of orchids!

Invasive species

There are many examples of plants and animals that humans have moved from one place to another, either deliberately or unknowingly. Native to Central and South America, *Vanilla* (*V. planifolia* and other species) was first cultivated by the Totonacs of eastern Mexico. In the 15th century, the Aztecs showed up, conquered the Totonacs, and discovered that the 'beans' tasted quite good. Off the plant went into the hands of the Spanish conquistadors, and the rest is history. Some have speculated that the trade of *Vanilla* species was so widespread among the indigenous peoples, that it eventually made its way into the wilds of southern Florida, taking root there. We assume that Florida's vanillas are native, but are they really? It is certainly plausible that indigenous people in the Americas played some role in transporting vanillas, just like their other valued crop plants, such as cacao, papaya, and guava.

Today there are consequences for the health of the planet's ecosystems from moving orchids by foot, car, boat or plane. Jonathan Swift summed it up best when he inspired the popular phrase, 'Fleas have fleas ad infinitum', to describe how small creatures are plagued by even smaller creatures. Do orchids have fleas? Yes, if you place scale insects and mealybugs into the same category: as parasites they sip the juices of plants, weakening their hosts. When we move infested orchids around, these tiny creatures infest healthy orchids in their new setting. Today in the Florida Everglades, native orchids are plagued by several different species of scale insect, all of which are exotic and are now invasive. In 2013, the orchid mealybug (*Pseudococcus microcirculus*), known as a ubiquitous pest of greenhouses in California, somehow made its way into south Florida and onto native epiphytic orchids, where it has become a pest. But, as Swift's phrase suggests, even scales and mealybugs have 'fleas' in the form of parasitic wasps that kill the offender, serving as biological control agents.

These examples are themselves a microcosm of a larger global problem involving exotic invasive species that humans have unleashed through travel. Orchids are also guilty invaders. Examples include *Polystachya concreta*, an epiphytic species now spreading into habitats in Hawaii, and Africa's *Oeceoclades maculata*, a terrestrial that is invading leaf litter throughout south Florida. In nearby Cuba, *Phaius tankervilleae*, originally from China, is now popping up in large numbers. As the world warms, invasive species will continue to spread and flourish, adding another layer of complexity to orchid conservation. It is probably no longer feasible to eradicate the invaders, so the best we can do is to control their numbers and limit their spread.

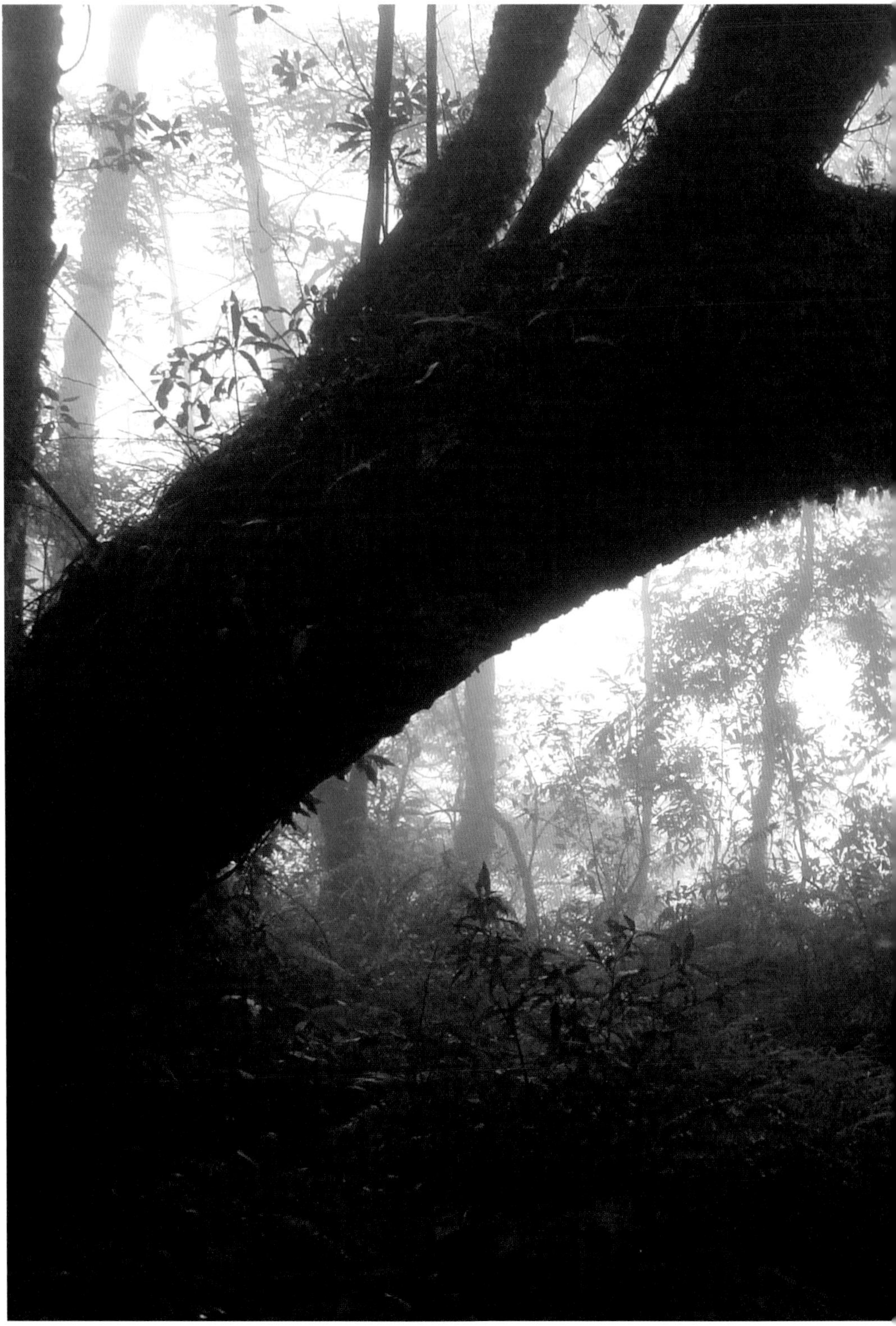

11 Who is doing what and where?

'There can be no purpose more inspiring than to begin the age of restoration, reweaving the wondrous diversity of life that still surrounds us.' E. O. Wilson

(previous pages)
Searching for orchids in the cloud forest of Taiwan.
Photo: Rebecca Hsu

A tale of two orchids
The story of *Laelia dawsonii* and *L. anceps* in Mexico encapsulates many of the interwoven and complex conservation issues described throughout this book. There are two forms of *L. dawsonii*: *L. dawsonii* f. *dawsonii* and *L. dawsonii* f. *chilapensis*, and they demonstrate the importance of involving local communities and of *ex situ* conservation. Since its discovery by orchid hunters in the 19th century, *L. dawsonii* has been subject to various degrees of taxonomic indecision. It was described as a distinct species in 1902, only to be later downgraded to the status of subspecies or a variety of *L. anceps*. Recently it has been restored to the status of a distinct species in its own right. Whatever their taxonomic status, both *L. dawsonii* and *L. anceps* are equally deserving of conservation, both in the wild and in cultivation.

First described by John Lindley in 1835 from plants imported into the UK by Loddiges & Sons, *Laelia anceps* rapidly became popular among growers due to the diversity in its flowers' colour, shape, and size. It was exported from Mexico in enormous quantities. Joseph Chamberlain alone had a thousand plants in his 'Mexican greenhouse' at Highbury in Birmingham, England. It is still one of the most widely cultivated species found in amateur orchid collections in both the UK and the USA. Numerous populations can still be found growing along the Sierra Madre Oriental that runs parallel to the Gulf of Mexico. There are two distinct varieties: *L. anceps* var. *anceps* grows in the state of Veracruz, where it is abundant and the source of many cultivated varieties, whereas *L. anceps* var. *mayensis* is found further south in Chiapas. Despite extensive clearance of its natural habitat, *L. anceps* is a survivor, and today thrives in the coffee plantations that have replaced the original cloud forest, for example.

By contrast, it appears that *Laelia dawsonii* has always been rare in nature. It was first introduced into cultivation in the British Isles by John Tucker in 1865. He found it growing in the Pacific region in the Sierra Madre del Sur. In a letter to his employers, Tucker wrote: 'The plants were found near Juquila, at a high elevation, about 150 miles from Oajaca [Oaxaca], in a *barranca* [ravine], growing in a rock, 2 to 3 feet from the ground, by the side of a running stream fully exposed to the sun – as there are no trees near – a very warm spot, sheltered on all sides by mountains. The locality is, however, remarkable for its extremes of temperature being very warm during day, and very cold during night.' He concludes by saying that although he had 'hunted' the whole neighbourhood, he could not find more than 60 plants.

Laelia dawsonii first flowered in the UK in the collection of Mr. T. Dawson. Initially it was thought to be a white form of *L. anceps*, and was considered to be one of the most beautiful introductions of the times, comparable with that of *Odontoglossum crispum*, or either one of the white moth orchids, *Phalaenopsis amabilis* and *P. grandiflora*. Today, two distinct geographical forms of *L. dawsonii* are recognised: the white-flowered *L. dawsonii* f. *dawsonii* and *L. dawsonii* f. *chilapensis*. So-called superior forms of *L. anceps*, with pale pink tepals and a velvety dark purple lip, which are sometimes offered for sale as *L. anceps* var. Guerrero, are in fact *L. dawsonii* f. *chilapensis*. *Laelia. dawsonii* f. *chilapensis* survives in cultivation in many villages around Chilapa in the state of Guerrero. Indeed, it was only known in cultivation until a small number of plants were discovered by Edgar Salmerón growing in the wild a few years ago. It is known locally as *calaverita* (little skull) and plays an important role in the local culture. Crosses are adorned with sprays of this orchid on el *Día de los Muertos*, the Day of the Dead (1–2 November).

A small number of plants of *Laelia dawsonii* f. *dawsonii* have been maintained under cultivation in a handful of towns in the state of Oaxaca. It is virtually extinct in the wild due to illegal collection. Federico Halbinger and Miguel Soto said that, in 1987, they found what was clearly a relict population of just 12 clones that was endangered by human disturbance, close to the place where Tucker made the original collection. Recently around 100 plants of *L. dawsonii* f. *dawsonii* were discovered for the first time growing in the state of Jalisco. The exact location of the plants remains secret, to guard against the possibility of illegal collection. Many people in Mexico still buy wild-collected plants despite the fact that they are contributing to the species' eventual demise in its natural habitat.

Today, *Laelia dawsonii* f. *dawsonii* is rarely found in collections in the UK. Why? Part of the answer probably lies in its rarity. Always rare in its natural habitat, early collectors were unable to discover more plants growing in the forests. Instead, all the plants they obtained had been cultivated and grown in trees in front of or near the huts of indigenous people, where they had been growing for centuries. All these plants had to be purchased separately. As the flowers were used for their religious ceremonies, the indigenous people were understandably reluctant to part with their specimens.

As *Laelia anceps* grew along the east coast, plants were easily transported to Veracruz, from where they were shipped to the UK. *Laelia dawsonii*, by contrast, grew along the Pacific coast, and their journey began with a voyage by sea along the coast to Panama by small steamer, with frequent stops at all intermediate ports. The plants then crossed the isthmus of Panama by rail (the Panama Canal didn't open until 1914), after which they were shipped to

England via steamers from Colón. Many plants did not survive the journey, and those that did were badly shrivelled, and took a long time to establish. Although *L. dawsonii* could not have been imported into the UK in such large quantities as *L. anceps*, it was regularly advertised for sale in *The Gardeners' Chronicle* in the 19th century. For example, in 1881 'an immense mass' of *L. dawsonii* was offered for sale at Stevens' auction rooms on behalf of Mr. F. Sander.

At that time, *Laelia dawsonii* f. *dawsonii* had a reputation for being more difficult to grow than *L. anceps*, although various authors disputed this, saying that it was simply a case of supplying the plants with more light and a little more heat than *L. anceps*. Indeed, in 1887, Ludwig Kienast-Zölly writes: 'I have often heard great complaints concerning the unsuccessful culture of the white varieties of *L. anceps* [i.e., *L. dawsonii* f. *dawsonii*], but which I find unjustified, as there is nothing more easy than to grow these plants as well as the type [i.e. *L. anceps*]; with the exception that the white-flowered forms require a temperature of 1° to 3°F [0.6° to 1.8°C] higher in winter as they come from the Pacific coast, which is warmer than that of the Gulf of Mexico.'

We should never underestimate what a small group of dedicated individuals can achieve. Although it can be very difficult to wean orchid collectors off their preference for wild-collected specimens, a group of orchid growers, including Sandro Cusi of Orquídeas Río Verde and Juan Morales from the Tepoztlán Orchid Society, are raising *Laelia dawsonii* f. *dawsonii* from seed, with the aim of using it as a symbol of Mexico's native orchids, and as a flagship for conservation. The collection and propagation effort at Orquídeas Río Verde began with just three clones, one originating in the USA, and the other two from growers in Mexico. Growers now stage a small show each autumn devoted exclusively to *L. dawsonii*. These shows have been successful in promoting conservation, and a good number of growers now buy flasks, having realised that orchids grown from seed can be outstanding, and it is not necessary to collect plants from the wild.

VIEWPOINT

The bumblebee orchid
by Andrés Ramos

In the cloud-covered heights of the forests of Veracruz in eastern Mexico, *Acineta barkeri*, colloquially known as *jicotera* (bumblebee's nest), *jicote*, and *boca de león* (lion's mouth), reveals a world of contrast and challenge. With its large, golden inflorescences exuding a sweet scent, it is more than a botanical marvel – it's a symbol of the fragile balance in its ecosystem. These forests, once rich in biodiversity, now face the encroachment of coffee plantations. The habitat that nurtures the orchid also produces some of the country's finest coffee, leading to a bitter trade-off between agriculture and the preservation of natural habitats.

The striking inflorescence of *Acineta barkeri* is reminiscent of a bumblebee's nest

(top)
***Laelia dawsonii*
f. *chilapensis*,**

as dozens of *Eulaema* bees can be seen hovering around it, or a lion's open maw into which these bees fly willingly. However, while drawn by the scent and eager to get closer, they are unable to fulfil their role as pollinators. This mismatch adds a layer of mystery to the orchid's existence and highlights the complexity of its survival in the wild. The rarity of *A. barkeri*, especially during its bloom in the summer months, makes it a coveted prize in the illegal orchid trade. Poachers, drawn to its beauty, often remove these plants from their natural habitat, unaware or indifferent to the delicate balance they disrupt. Once removed, the survival of the orchid is unlikely, given its specific needs for high humidity and a unique symbiotic relationship with surrounding epiphytes and humus.

At the conservation charity Orchidarc, our mission is to turn the tide for *Acineta barkeri*. Collaborating with Universidad Veracruzana, we've embarked on a journey to pollinate and grow these orchids from seed, a step crucial for both understanding and conserving the species. The process is fraught with challenges, not least of which is the orchid's self-incompatibility and complex pollination requirements. Despite these hurdles, we have achieved modest success in artificial pollination, yielding a few pods [seed capsules] per year that provide a glimmer of hope in our conservation efforts.

The answer to the mystery of *Acineta barkeri's* true pollinator continues to elude us, hidden within the dense foliage of its native cloud forests. Observations suggest that the pollination mechanism may involve a yet-unknown thermal trigger, an intriguing possibility that demands further study. This elusive mechanism, vital for the natural propagation of *A. barkeri*, is a key piece in the puzzle of its survival. As we delve deeper into the mysteries of this magnificent species, our efforts not only aim to preserve the few thousand specimens left, but also to reintroduce them into areas where they have been nearly eradicated, restoring the natural splendour of their cloud forest home.

Acineta barkeri, *Prosthechea vitellina* and bromeliads for sale in Mexico.
Photo: Orchidarc

Epiphyte biologists climb trees
'But all is out of reach of the curious and admiring naturalist. It is only over the outside of the great dome of verdure exposed to the vertical rays of the sun that flowers are produced, and on many of these trees there is not a single blossom to be found at a height less than a hundred feet. The whole glory of these forests could only be seen by gently sailing in a balloon over the undulating flowery surface above: such a treat is perhaps reserved for the traveller of a future age.' Alfred Russel Wallace, 1848

More than a century after Wallace wrote this, his dream of being able to study plants high in the forest canopy has become a reality. Techniques for accessing the canopy include canopy cranes, aerial walkways, gondolas suspended from cables, balloons, and drones. But surely the most exciting way to study the orchids is to climb the trees.

The ancient coniferous cloud forests of Taiwan
Imagine climbing 200 ft (61 m) to stand at the top of an ancient giant conifer. Emerging through the canopy, high above its neighbours in the cloud forests in the mountains of Taiwan, *Taiwania cryptomerioides* is one of the tallest trees in Asia. Few studies have been carried out in the canopies of these old-growth trees, and little is known about their associated arboreal communities, or the structural complexity that influences their distribution and composition.

In 2014, Rebecca Hsu organised a team of six to climb the legendary group of giant taiwanias known as the 'Three Sisters' that grow in the primary forests of north-eastern Taiwan. Despite enduring periods of devastating logging, when timber extraction took place on a massive scale, Taiwan has retained large areas of old-growth forests. Although the island is small, Taiwan has over 200 mountain peaks greater than 3,000 m (9,800 ft) in elevation. About 20% of the vegetation is cloud forest. The dominant canopy trees of these forests include Taiwan cypress (*Chamaecyparis formosensis* and *C. obtusa* var. *formosana*), Taiwanese firs, Taiwan spruce (*Picea morrisonicola*), *Rhododendron formosanum*, *Pinus armandii* var. *mastersiana*, and tropical hardwoods.

After logging was banned in 1989, the access roads that penetrated deep into mountains soon became overgrown due to damage caused by the frequent typhoons and lush tropical plant growth. Driving along these roads, time and again the team was forced to stop and use chainsaws to remove fallen trees that were blocking the road. After spending a night at about 2,000 m (6,500 ft), camping under an umbrella-shaped Taiwan red pine (*Pinus taiwanensis*), the following morning, they packed up their tree-climbing gear and, crossing numerous landslides and streams, drove the final five kilometres to their target, three giant Taiwanese firs that had survived the logging era.

(left)
Rebecca Hsu and her husband Brian Chiu on the tallest camphor tree in the world in the cloud forests of Taiwan.
Photo: Mr "Luo"

(opposite top)
Epigeneium fargesii **carpeting the branches of a** *Taiwania*.
Photo: Mr Shen Kun Yu

(opposite bottom)
Pleione formosana **growing on** *Chamaecyparis obtusa* **var.** *formosana*. *Pleione formosana* **is illegally harvested in Taiwan.**
Photo: Rebecca Hsu

169 WHO IS DOING WHAT AND WHERE?

They ascended the Second Sister by rope, its enormous limbs cloaked in thick carpets of epiphytes, including lichens, orchids, ferns, and rhododendrons (*Rhododendron kawakamii*). The entire tree was laden with *Davallia clarkei*, a common cloud-forest fern, and – higher in the canopy – with another fern, *Phymatopteris quasidivaricata*. There were two epiphytic orchids, *Epigeneium fargesii* and *Dendrobium moniliforme*. The population of *E. fargesii* densely carpeted the branches. Black scars on the tree trunks told the story of frequent lightning strikes originating from the towering clouds that drift over the island. Such scars on behemoth trees in Taiwan and throughout the tropics serve as an ominous reminder of some of the dangers faced by those who are willing to ascend the canopy to study epiphytes.

Araucaria angustifolia with *Sophronitis* (*Cattleya*) *coccinea* in the south of Brazil.
Photo: Stig Dalström

Atlantic Forest, Brazil

Mention the Brazilian rainforest, and deforestation of the Amazon immediately springs to mind. We should of course be concerned as parts of the Amazon may be approaching a tipping point, when the forest will be replaced by savannah, leading to potentially disastrous consequences for the global climate. What is less well-known is that there is another rainforest in Brazil, the Atlantic Forest. Once stretching over 1.5 million km^2 (about 600 thousand mi²) along almost the entire east coast of Brazil into parts of Paraguay, Uruguay, and Argentina, today little more than 7% of the original cover remains. Centuries of deforestation for timber, sugar cane, coffee, cattle ranching, and urban sprawl are to blame for its demise.

More ancient than its famous Amazon neighbour to the north, the Atlantic Forest is home to 60% of Brazil's biodiversity, and is one of the world's top biodiversity hotspots. The reason for this incredible diversity may be the forest's great length. Running along the South American Atlantic coast, it spans both tropical and subtropical latitudes. Adding to the richness is a highly heterogeneous landscape that supports many different vegetation types, including semi-deciduous forests that drop a portion of their leaves during dry spells. The trees in the various vegetation types are colonised by many different orchids and other epiphytic plants, contributing to the region's high tally of endemic species. As with other hotspots, Brazil's Atlantic Forest – or what's left of it – remains vulnerable to human activity exacerbated by climate change, and many of its unique species remain in peril.

Brazil's Atlantic Forest comprises three distinct forest types: coastal forest or Atlantic rainforest (*Mata Atlántica*); Atlantic semi-deciduous forest; and Araucaria mixed forest. To gain a better understanding of their species diversity, the Atlantic Forest's Semi-deciduous biome in Cantareira contains around 197 different tree species in an area of just 74 hectares (183 acres). By comparison, the entire UK has around 50 native tree species. The Araucaria mixed forest is doubly significant as Araucarias are gymnosperms that evolved just after the Earth's greatest extinction event at the end of the Permian geological period, 251 million years ago, which set the stage for the Age of Dinosaurs. It is not hard to imagine seeing giant sauropods with their long necks feasting on the upper branches of these once majestic and strange trees. Unfortunately, most people view forests like this rather differently. *Araucaria angustifolia*, known as Paraná pine, was cut down and exported in huge quantities in the 20th century. As a result, it is now critically endangered. What little remains of these forests are home to such jewels as *Cattleya coccinea* (syn. *Sophronitis coccinea*), and other epiphytic orchids that are found nowhere else on earth.

More than 35 years at Macaé de Cima, Brazil
An analysis recently published by the conservation project, 'A Trillion Trees' highlighted the Atlantic Forest of Brazil, where a project to restore the forest has regrown an area of 4.3 million hectares (11 million acres) since 2000. As part of this valuable work, David and Izabel Miller, botanist Richard Warren and the Rio Atlantic Forest Trust (RAFT) have aimed, not only to restore forest, but also to document the progress of regeneration over a period of 40 years.

VIEWPOINT

Cantareira
by Luciano Zandoná

The Serra da Cantareira mountain range is the domain of dense montane rainforest. Out of a total 32,000 hectares (79,000 acres), 8,000 hectares (19,768 acres) are protected as an 'integral protection conservation unit' in Cantareira State Park, near São Paulo, Brazil. The moist forest can reach more than 2,500 m (8,200 ft) into the rocky (rupestrian) fields – high-altitude fields that are part of a very specific type of upper montane forest that occurs at altitudes above 1,500 m (4,900 ft).

Cantareira is the forest of my heart, where I was born, raised, and live today. Here I grew up in contact with nature, learning to respect it. It is my intention to continue working for the conservation of this forest, while I'm here, in this life. I was born in 1975, and have seen a lot of forest being cut down in my lifetime, and I still see it today. Forests that are not protected as conservation units are disappearing, even here in Cantareira. The extent of land grabbing in the forest on the edges of Cantareira State Park today is frightening. There is an urgent need to restore forests, but we are far from it.

Most of the many orchid species occur in small populations and have been suffering as a result of climatic change and the edge effect – damaging factors that penetrate the forest such as the desiccating effects of warmer air, increases in light that promote the growth of weeds, invasive species, decline of pollinators, and illegal collection. When I began studying orchids in the Atlantic Forest in 2008, I realised that there were two ways to study epiphytes, most of which grow on large trees in different forest strata. The first was to evaluate large fallen trees. The other way to study epiphyte communities and their interactions was to access the canopy using climbing equipment.

Heavy summer rains in the region saturate both the soil and increase the weight of the epiphyte gardens high in the canopy, often leading to forest giants crashing to the earth. Our work is based on obtaining material from fallen trees, with a small portion going to botanical collections. During the wet season, most of the rescued orchids are relocated in their original habitat. Climbing has become the most efficient way of reintroducing the plants: it allows us to access different heights and locations on the trees, enabling us to distribute them in conditions closer to their original habitat, leading to greater success in terms of survival and establishment. Three years ago, for example, I successfully rescued and relocated 100 *Oncidium varicosum* orchids in a big pink cedar (*Cedrela fissilis*) and followed them until they flowered.

After selecting the tree to be climbed, we use a hand slingshot to throw a lead weight with a thin nylon thread attached over the branches we wish to access. We then pass a thin rope that is used as a guide to pull the main rope for the canopy climb. Anchoring to branches has been my biggest concern before climbing a tree. To ensure safety, the tree is first inspected for the presence of rotten branches or evidence of termites. I always run the guide rope over two branches at the very least, and sometimes I cross the entire crown with the rope. The higher the rope is, the greater is my mobility through the tree.

Animals always demand our attention – after all, the tree is their home. The presence of bees and wasps can be a problem, in which case a bee keeper's outfit is worn. The ideal strategy is always to climb away from the main trunk of the tree, thereby preventing any contact with ants and other stinging or biting insects. After the main rope is passed and anchored, comes the good part: climbing the tree and evaluating the plants up there. Making use of ascension techniques, you reach different heights in the treetops. Canopy work has many applications besides studying orchids, bromeliads, and a multitude of other plants. I am able to take photographs of the flowering species, in their habitat, sometimes getting a glimpse of interactions

(above)
Luciano Zandoná and *Oncidium varicosum* **in the canopy in Cantareira, Brazil.**
Photo: Luciano Zandoná

(below)
Stanhopea lietzei **and a bee pollinator.**
Photo: Luciano Zandoná

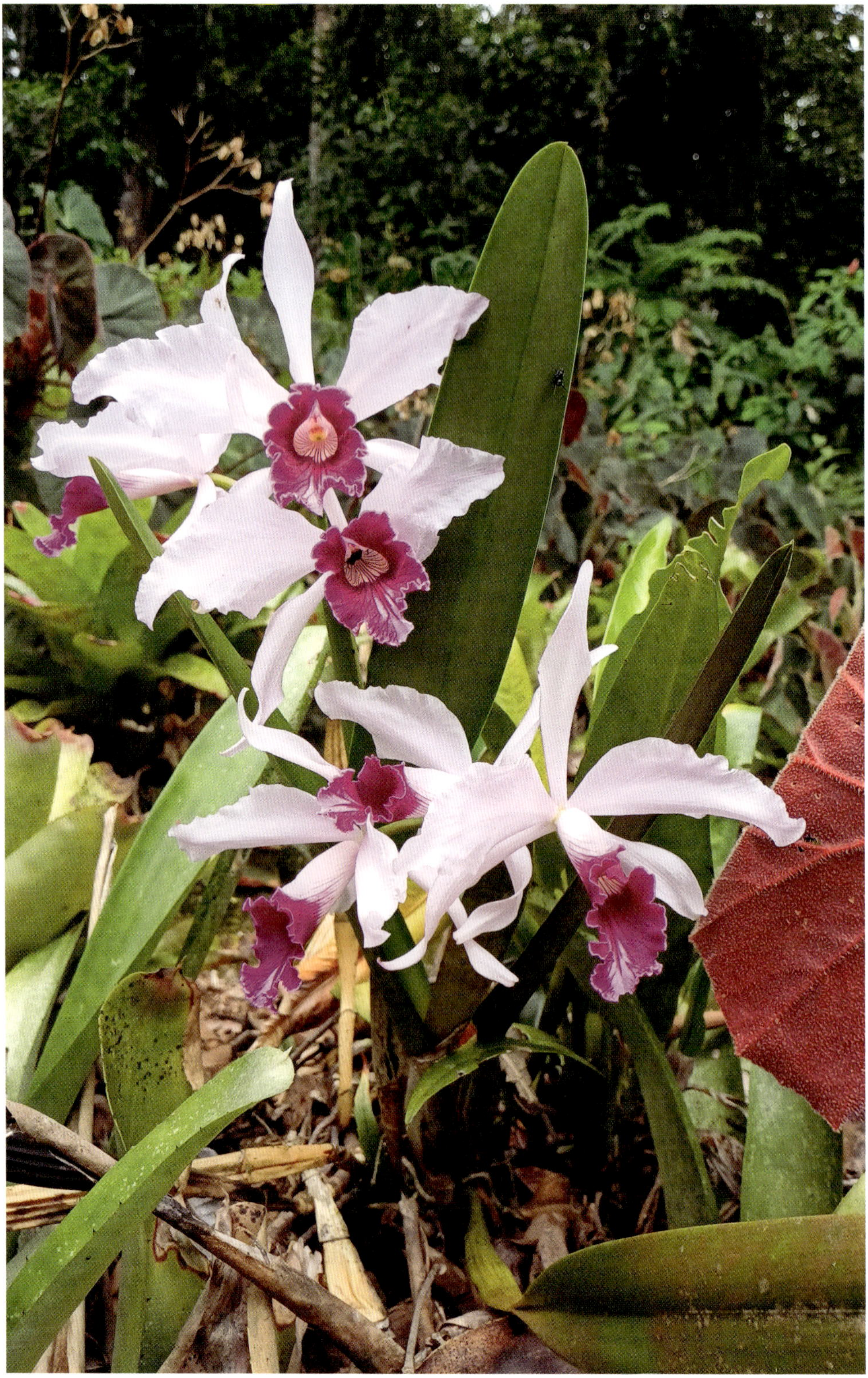

with pollinators. On one occasion I climbed a large fig tree to photograph *Stanhopea lietzei* in bloom, and was pleasantly surprised to be able to observe and photograph a *Bombus* bee visiting the flowers. Each tree is a world of possibilities for new discoveries, especially when it comes to micro-orchids, because even with powerful binoculars it is nearly impossible to discover what exists on the trees without actually climbing them.

The Guarani, indigenous people who live in the Atlantic Forest, call *Cattleya purpurata* 'donkey's ear' because, when the plants grow in the shade, their leaves become large and floppy – like donkeys' ears. When grown in bright light, the leaves are stiff and erect. *Cattleya purpurata* is endemic to the Atlantic Forest of south-eastern Brazil, where it is found at low elevations from the restinga to the hillside forests along the rivers, up to about 100 metres altitude. It can live in the canopy of trees both large and small, and also on rocks as a lithophyte. The plants illustrated were rescued about two years ago on a project where native vegetation was cut to provide a corridor of more than 20 kilometres through the forest for the construction of electricity pylons. A total of 26 plants were rescued and relocated. The plants in the photograph were placed on a rock at the foot of the mountain to allow their seeds to disperse into the surrounding area. Already there is a fruit that was sown in February 2023 and there are many seedlings being grown *in vitro* ready for reintroduction.

In the state of São Paulo, a government licence is needed to build anything that requires the cutting down of forest. One of the conditions of the licence is that steps must be taken to rescue the vegetation before the work begins. We collect everything possible and relocate it to nearby areas with the same environmental conditions – generally in protected areas such as conservation units. In this project the plants were relocated to the state park in Serra do Mar, which is home to the largest continuous Atlantic Forest in Brazil.

Cattleya purpurata established on a rock and in full bloom in Brazil.
Photo: Luciano Zandoná

In the early 1970s, David Miller bought 120 hectares (300 acres) of mostly original forest in Macaé de Cima in the Serra do Mar in Rio de Janeiro state. His mantra was: 'If you are going to conserve, you must know what you are conserving.' To achieve this goal, Miller and Warren planned a two-pronged attack. First, they spent several years work with staff and students at the Rio de Janeiro Botanic Garden to conduct a thorough survey of the original forest. In one hectare of forest, they identified over 3,000 trees belonging to over 300 species. Three of these species were new to science. Many specialists from other disciplines have also visited and worked in the reserve, and as a result, there is now an inventory of all birds documented in the area, for example.

Second, they studied the orchid family, and over 35 years documented and photographed over 600 species growing in the Organ Mountain Range. Not only were these plants described *in situ*, many were also hand-pollinated, the seed germinated, and seedlings made available to growers along with accurate habitat and climate information. For more than 40 years, they also monitored regenerating forest that had been destroyed by fire. They identified three vital factors that prevent regeneration: further fires, (usually recreational); domestic animals that will graze on any green matter during the dry winter months; and human activity, since even early regrowth scrub can be used as fuel or fencing and building materials. If these factors are kept at bay, early growth – even on nutrient-poor soils – can occur, gradually building up humus. A small range of shrubs appear, mainly Asteraceae and Melastomataceae species, which have wind-borne seeds. Colonisation slowly proceeds. After about 25 years, these shrubs have grown and bloomed, and the fruits and flowers are visited by birds, bats and other animals from adjoining original forest. Their droppings contain seeds of a fresh selection of species, such as palms and more noble trees.

In parallel with tree regrowth, Miller and Warren also studied the orchids that colonised the area. At the start, because the canopy had not yet closed over, humidity was low, and light levels high. Not only terrestrial orchids grew on the ground; even epiphytic species such as encyclias, epidendrums, bifrenarias, and oncidiums grew alongside bromeliads on the forest floor. After 20 years or so, when the canopy had closed and the humidity had risen, orchids, water-bearing bromeliads, mosses, and ferns were found growing as epiphytes.

A thorough survey of the area after 45 years revealed a healthy secondary forest with trees heavily colonised by bromeliads, ferns, and about 25 species of orchids. Considering that the nearby original forest contains at least 230 orchid species, and the prevailing wind blows towards the regrowth, the paucity of orchid species colonising the regrowth is puzzling. Clearly the species found are pioneers, but what are the qualities that these pioneers have that enables them to succeed? Miller and Warren artificially reintroduced orchids with the twin aims of re-establishing plants taken from deforested areas and enhancing populations in areas where plants had been over-collected in the past. In addition, they wanted to see how the reintroduced plants reacted and adapted to conditions in areas of forest at different stages of maturity.

In 1976, they had the opportunity to investigate the possibility of reintroducing plants of *Laelia crispa* into a section of disturbed forest at between 1,200 and 1,400 m (3,900 and 4,600 ft). Subsistence farmers had cut down around two hectares (4.9 acres) of original forest, and Miller and Warren found that most of the felled trees had colonies of *L. crispa* growing high in the forks of the trees. They carried hundreds of plants back to their reserve at Macaé de Cima. Even so, thousands were left behind to either rot or be burnt.

The plants were fixed low down onto the trunks of trees in suitable light and wind conditions. The vast majority flourished and flowered well. The flowers were visited by hummingbirds, euglossine, bumble and carpenter bees, and produced seed without the necessity for hand pollination. Seedlings appeared in profusion both above and below the parent plants, and on neighbouring trees.

Most of the seedlings died, but the first seedling flowered ten years later. Incidentally, the grouping of seedlings around the parent plant accounts for the finding that the plants tend to grow in large colonies, making it easier for plant collectors to locate specimens.

In contrast to their experience in disturbed but original forest, relocation into regrowth forest was more challenging. If attached to normal, healthy young trees, the plants generally failed to thrive, producing only a few roots, small shoots and no flowers. If, however, the plants were attached low enough on the trunk so that the new roots soon encountered the soil, then the plants grew and flowered swiftly, producing seed after natural pollination. One particular colony was monitored for over 22 years. Each year it probably produced millions of seeds that floated over the surrounding moss-covered bark and humus. Not a single seedling was found. They hypothesised that either there was something missing, or something inhibitory was present.

In recognition of this successful conservation work, Richard Warren was awarded the UK's prestigious Westonbirt Orchid Medal in 2014. The State Environment Institute gave the Millers' land permanent protection as a Private Natural Heritage Reserve, which demonstrates how much can be achieved by enthusiastic yet voluntary work. The reserve is now linked to the larger Three Peaks Reserve.

Building on the truly inspiring work carried out by David and Izabel Miller and Richard Warren over many years, Alexandre Antonelli, Director of Science at Kew, and his wife Anna, recently acquired the reserve, previously known as Sitio Bacchus. It will form the hub of an initiative called ARAÇÁ (the Atlantic Forest Research and Conservation Alliance). The aim is to unlock knowledge in a megadiverse nucleus of Brazil's Atlantic Rainforest by promoting collaborations, fostering discovery-driven research and conservation, and catalysing training and education. ARAÇÁ will establish partnerships with local and international collaborators and researchers to collect baseline knowledge of all biodiversity on the site and to conduct short- and long-term research projects, monitoring programmes, and conservation initiatives. Antonelli and his colleagues at Kew, Campinas University in São Paulo, and others, will use their expertise to ensure that the reserve continues to be an important centre for orchid research and conservation, providing the foundations for further work in evolutionary biology, ecological interactions, and conservation biology.

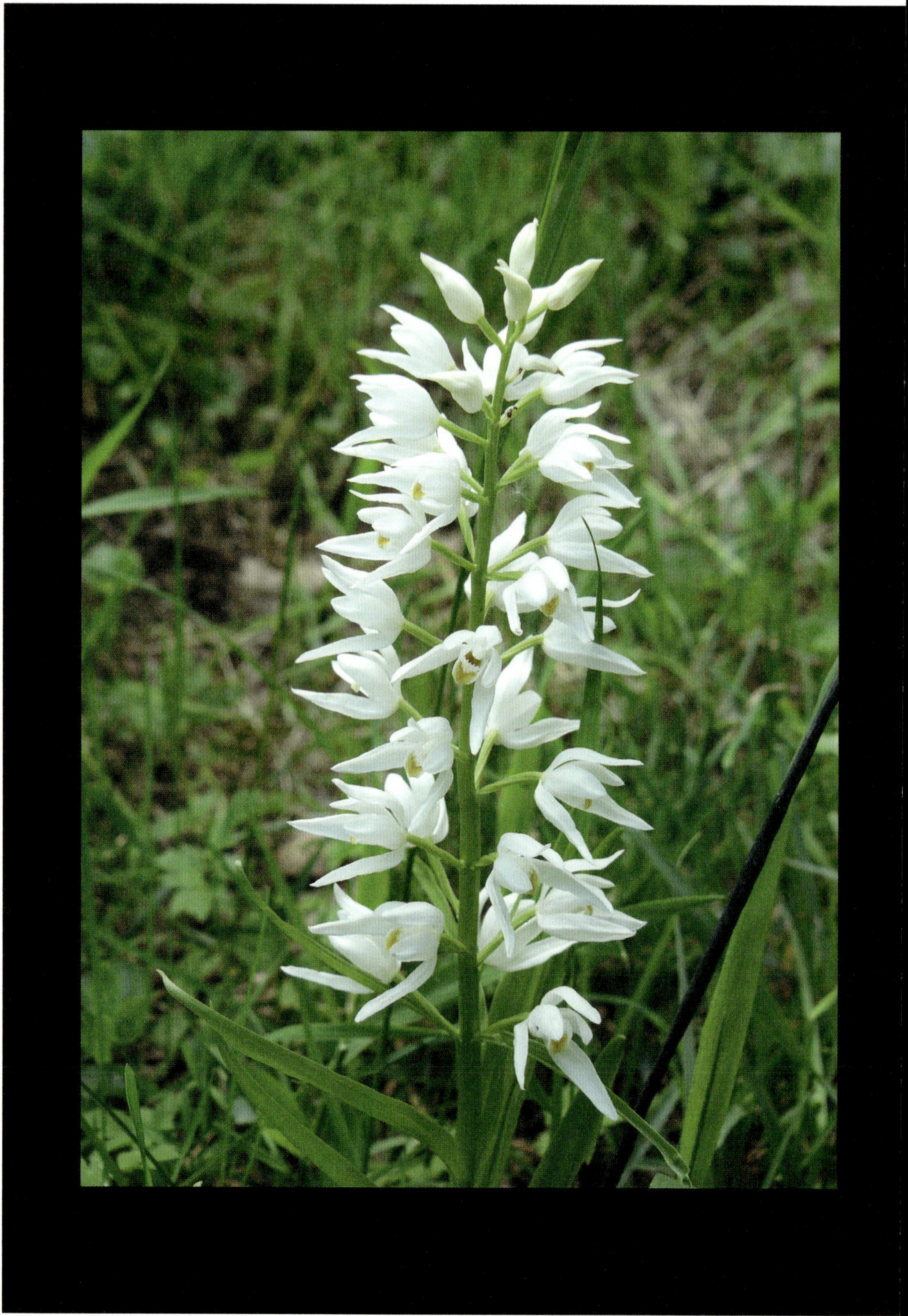

12 Where do we go from here?

'Want of foresight, unwillingness to act when action would be simple and effective . . . the features that constitute the endless repetition of history.' Winston Churchill, 1933

If we are serious about saving our orchids in a warming world, we have a lot of work to do. As temperatures rise, as the clock ticks, and as resources (natural and monetary) dwindle, we face difficult choices, knowing full well that we can't save every species. As conservation areas and reserves shrink in size over time, species are lost, with the rarer species within small, isolated areas tending to disappear first. But, as Kingsley Dixon has stated in his essay (see page 84), 'We can and will turn orchid conservation around.' The first step is to prioritise, to focus on the species that are most vulnerable, and especially endemic species.

Establishing conservation priorities
Where are the highest concentrations of endemics? One place to look is on islands. In the Caribbean hotspot, for example, *Dendrophylax fawcettii* and *Myrmecophila thomsoniana* (the banana orchid) are known to occur only on tiny Grand Cayman – both species constitute a conservation priority. Larger islands such as Madagascar and New Guinea are clearly priorities, considering that more than 85% of their orchid species are endemic. Although we think of islands as isolated patches of land surrounded by water, islands of vegetation also occur within large, land-locked areas on six of the world's seven continents. By the end of this century, it is not inconceivable that green plants will colonise vast areas of Antarctica as the ice caps recede. According to the British Antarctic Survey, less than 1% of the continent is currently suitable for plants, and only two flowering species are known to occur there naturally, one being a grass (*Deschampsia antarctica*) and the other a pearlwort (*Colobanthus quitensis*). Most of the current flora is dominated by bryophytes (mosses and liverworts) and fungi, including those found in lichens. It seems inevitable that, perhaps somewhere on Patagonia or Tasmania, seeds of a terrestrial orchid will make their way onto Antarctic soil, where the embryo will just happen to meet a suitable fungus to initiate germination on a balmy sunny day. Soon, other plants with heavier seeds will follow, eventually yielding islands of green vegetation.

Individual mountains can also be considered to be islands, when they are stranded in a landscape altered by human activity, or are high enough in elevation to be surrounded by clouds and the life-giving moisture they bring. Where the landscape is flat, green islands of vegetation – often square or rectangular in shape – are evident from above in areas surrounded by agricultural fields, parking lots, and other disturbances inflicted by our species. In countries that set aside forests in preserves such as national parks, these green islands are safe from encroachment so long as there are

(previous left hand page)
Understanding orchid populations requires long-term studies. The long-leafed helleborine, *Cephalanthera longifolia*, is rare in Britain, although it is common in parts of continental Europe.
Photo: Philip Seaton

(previous right hand page)
Populations of the green-winged orchid, *Anacamptis morio*, are declining in parts of the UK.
Photo: Philip Seaton

laws in place to safeguard these areas from development. But laws mean little when they cannot be enforced because of corruption and other factors linked to money or lack thereof. People in local communities, however, do occasionally band together for the common good of preserving their adjacent forests, as has happened in Madagascar, for example (see page 75). We need more tree-huggers.

While the focus tends to be on saving individual species, we must also be mindful of conserving higher taxonomic ranks such as genera, tribes, and subfamilies. As Ken Cameron points out, conservation is not just about richness as measured by numbers, but also about preserving diversity across the tree of life. Of the five orchid lineages, three are represented by relatively few species, and conserving species within these groups is a relatively achievable target.

Growing orchids from seed

'To a mycologist an orchid seedling without its fungus is like Hamlet without the Prince of Denmark.' John Ramsbottom, 1922

The first orchid seedlings were probably raised at the National Botanic Gardens, Glasnevin, Dublin, around 1844. By 1907, M. Lucien Linden's nursery in Brussels was home to more than a thousand *Odontoglossum crispum* seedlings. Although it was known as early as 1847 that orchid roots were normally infected with mycorrhizal fungi, and that seed would only germinate on compost in which an orchid of the same genus or species was being grown, it wasn't until the end of 19th century that Noël Bernard discovered the true importance of these fungi. He revealed that, in nature, every orchid seed's embryo must be infected by a specific mycorrhizal fungus to germinate. Bernard demonstrated this experimentally by isolating and using fungi from the roots of seedlings that had been sent to him by orchid growers, and from the roots of plants collected in the wild. A photograph published in *The Orchid Review* of 1906 shows *Odontoglossum* seedlings germinating on a 'nutrient jelly' slope, in a test-tube closed by a cotton-wool plug covered with a tinfoil cap, after infection with a fungus that had been 'found on the roots of odontoglossums'.

Around the same time, nurseryman Joseph Charlesworth was becoming increasingly dissatisfied with the relatively unproductive early method of sowing seed on compost in pots. After meeting John Ramsbottom, a mycologist at the British Museum of Natural History, he decided to conduct experiments of his own. The fruits of his endeavours can be seen in a photograph of one of Charlesworth's glasshouses, heavily shaded, with row upon row of conical flasks with cotton wool plugs, lying on their sides in racks. The legend to the photograph tells us that, 'The glass flasks are partially filled with a suitable compost, and after being sterilised by heat the necessary fungus is added. A few weeks later the seed is sown and germinates rapidly with considerable regularity.'

In 1922, Lewis Knudson revolutionised the raising of orchids from seed by demonstrating that many species will germinate without the need for the fungus, on media containing simple sugars and minerals. Suddenly it became relatively easy to grow many orchids from seed – particularly tropical species – using simple laboratory techniques. By 1928, Knudson's methods had largely replaced the symbiotic technique. However, it is worth noting that in 1956, David Sander reported that, despite the vast strides that had been made with asymbiotic culture, he still found that he could obtain more robust and quicker-growing plantlets if the seed had been sown on 'well-rooted parent plants', when compared with seedlings raised from the same seed capsule in the lab.

Debate about whether symbiotically raised seedlings are more vigorous than their asymbiotically raised counterparts continues today. In 1906, J.M. Black wondered if it would one day be possible for scientists to cultivate these fungi and distribute them to growers as an inoculum for pans of seedlings. Today, a revival of interest in growing orchids from seed using mycorrhizal fungi for reintroduction has led to such a sharing of fungi among amateur and professional growers of terrestrial orchids.

Problem children
With so many different kinds of orchids being grown from seed worldwide, it is easy to forget that some have resisted all attempts at their propagation. These 'problem children' may never be tamed in captivity. So-called mycotrophic orchids, for example, such as the achlorophyllous bird's-nest orchid, will probably frustrate efforts to grow them from seed indefinitely. With a little luck and perseverance, others may eventually reveal their propagation secrets. Among the photosynthesisers, temperate terrestrials are generally regarded as more difficult to grow from seed when compared with tropical epiphytes. Seeds of species growing in colder climes often require dormancy mechanisms to be broken, using cold and moisture treatments (stratification), for example, before they will germinate. The orchids from North America's tall-grass prairies, *Platanthera leucophaea* and *P. praeclara*, both require seeds to be moistened and chilled in darkness for months before they germinate appreciably. In nature, seeds of these and other species (*Cypripedium*, for example) may remain in the soil as a seed bank for a number of years.

Other photosynthetic terrestrials have resisted most attempts to cultivate them. They include the sword-leaved helleborine (*Cephalanthera longifolia*), *Isotria medeoloides*, and Hawaii's *Peristylus holochila*. Challenges remain, and there may soon come a day when specialists must abandon all hope of propagating an endangered 'problem child', simply because there are so many less

Catalina Restrepo holding a flask of seedlings at EAFIT University, Medellín, Colombia.
Photo: Philip Seaton

fastidious orchid species in need of their attention and resources.

Have we identified which species and habitats are most at risk? Not yet. Do they correspond with the already established hotspots for other species? We still don't know. There remain large gaps in our knowledge about the status of many species in the wild, and currently only a small proportion of orchids have been assessed for Red Listing. Once we have identified those species that are currently at risk, should we not only prioritise the preservation of their habitat, but also bring samples into cultivation in secure locations? Probably. Are orchids being found outside or inside reserves? Are these orchids ephemeral or are individual plants capable of living for decades? It depends. *Habenaria repens* lives as an aquatic orchid in Florida's swamps for only about a year and therefore must continuously set seed, whereas *Dendrophylax lindenii* may live for decades. There is still much we don't know. Do the new discoveries simply reflect where today's botanists live or the places where they choose to visit? Are they visiting the most

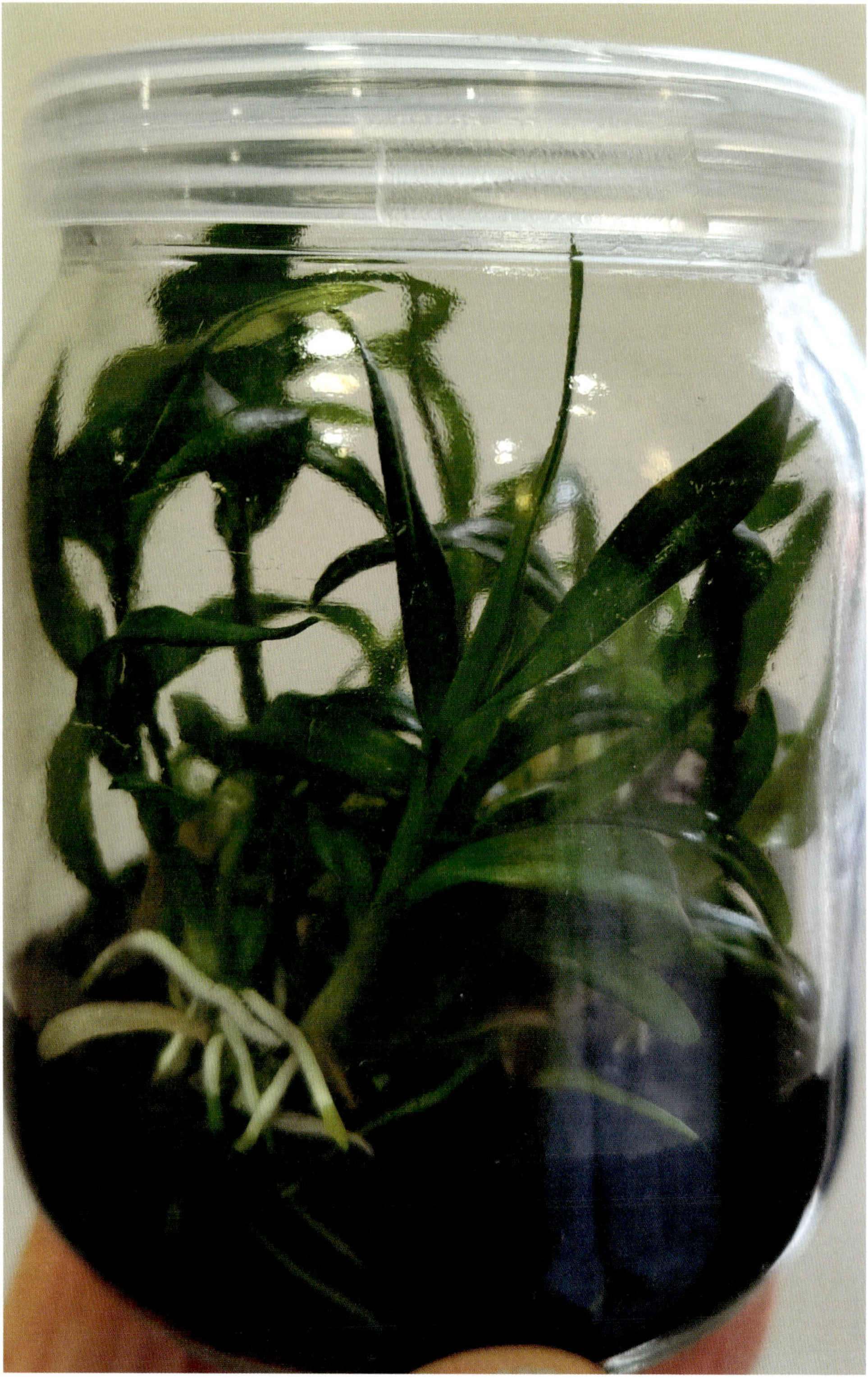

promising locations? To which genera do the recent discoveries belong – are they mostly pleurothallids in the Americas, for example?

Population studies of European terrestrial orchids
Nothing stays the same. If abandoned, the species composition of an English meadow will gradually change as the meadow is first colonised by shrubs and then trees, and eventually becomes woodland in a process known as ecological succession. Likewise, if a meadow is not cut annually for hay, and the hay removed, the species composition will change as nutrients accumulate and more vigorous species out-compete the less vigorous species, including the orchids.

In 1979, Terry Wells developed a simple triangulation method using wooden pegs, cup hooks, and measuring tape to monitor the fates of individual orchid plants over a number of years. It had been widely assumed that the bee orchid (*Ophrys apifera*) was both short-lived, and monocarpic – the plants flowered once and then died – but Wells demonstrated that they may live for up to 11 years. Annual records demonstrated that individual plants can flower for several years in succession, or miss out some years – not appearing above ground for one or two years, only to appear again the following year. By contrast, Michael Hutchings has shown that the related early spider orchid (*Ophrys sphegodes*) is relatively short-lived, with an average lifespan of just over two years. Long-term studies by other researchers suggest a maximum lifespan of 30 years for *Dactylorhiza sambucina*, 25 years for *D. incarnata* and 28 years for *Listera ovata*.

Orchid populations may be affected by several factors. Milder winters, or less frequent cold winters, may benefit some northern European species. Prolonged summer droughts may have an adverse effect, as can floods that saturate the ground, mechanical damage (trampling), and predation. Herbivores such as sheep and cattle, for example, graze on orchids before the plants set seed. Long-term studies provide information needed for the conservation of endangered species, enabling researchers to make informed predictions about population trends using mathematical tools to analyse their age structure, and computer software to generate models of what is likely to happen under different management regimes. What role will artificial intelligence (AI) play? How might it impact orchid conservation in the none-too-distant future? Only time will tell.

Flask of seedlings of
Dendrobium bigibbum.
Photo: Philip Seaton

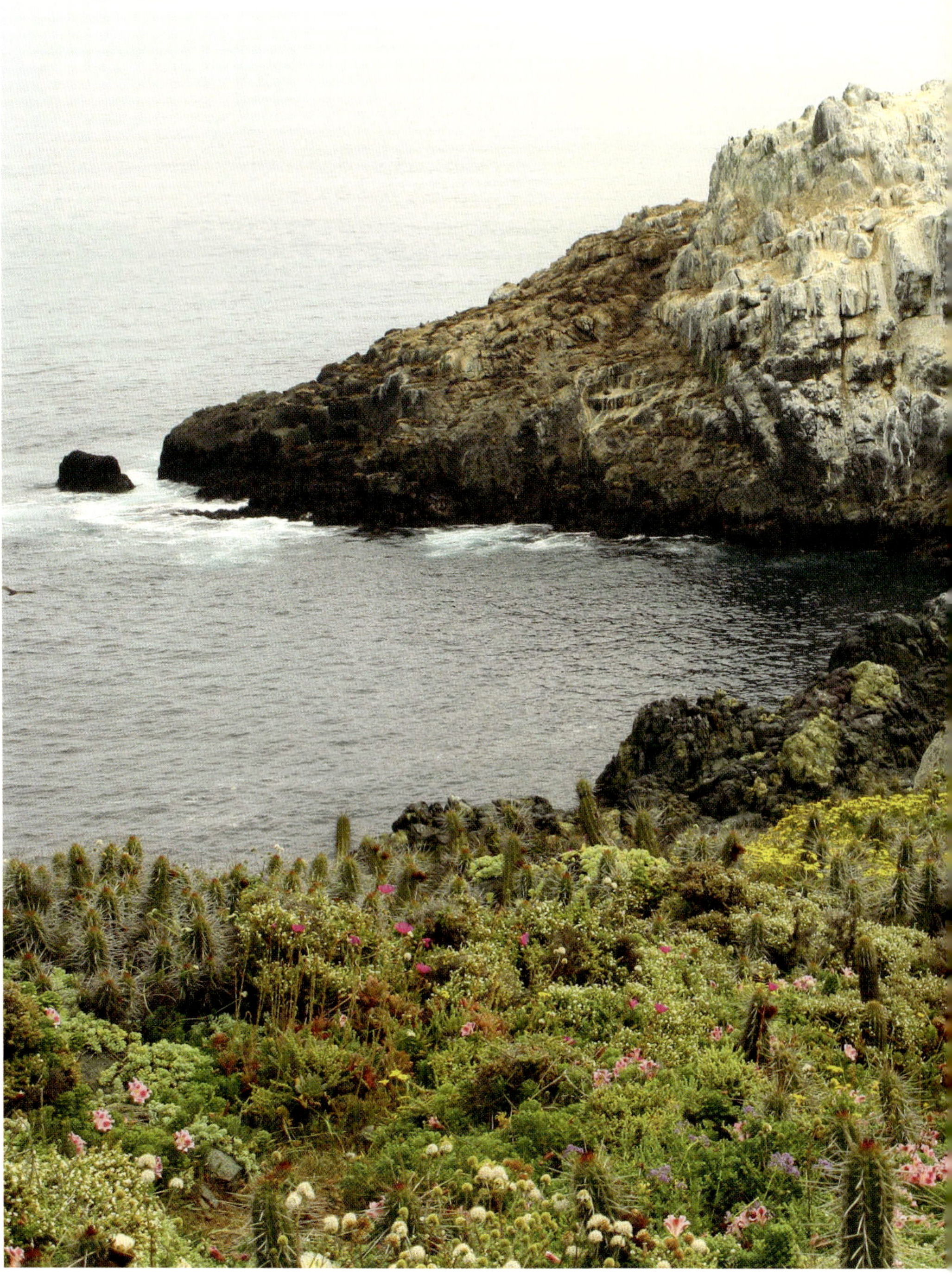

13 Education: the teachers and the storytellers

'Education is the most powerful weapon which you can use to change the world.' Nelson Mandela, 1990

(previous pages)
Schoolchildren participating in a spring expedition to the Santuario de Los Molles, Chile.
Photo: Sergio Elórtegui Francioli

As people move from the countryside into the urban environment, they increasingly become disconnected from the natural world. This disconnect is exacerbated in young people by the unprecedented pace and dominance of technological change. The indoor and the virtual are replacing the outdoor and the natural. There is a danger that nature is merely seen as media entertainment. The very words describing the natural world are being lost. In the latest edition of the *Oxford Junior Dictionary*, words such as *acorn*, *buttercup*, and *willow*, have been replaced with words such as *blog*, *celebrity*, and *voicemail* that are said to better reflect the experience of modern-day childhood, in which fewer children live in semi-rural environments and experience the changes of the seasons. Education in most countries focuses on completing the curriculum and on examination results. Children and their teachers find themselves confined to the classroom, with little time devoted to outdoor studies. If they do venture outdoors, they often take their technology and internet with them, adding another barrier between them and the natural world.

And then there is the phenomenon of 'plant blindness', a term coined 20 years ago by two botanists in the USA. On a personal level, we don't notice plants in our environment, perhaps because they all appear green to our eyes, and blend into one solid mass. It is hard to believe that 'plant blindness' was rampant before there were pharmacies and supermarkets, considering that our gardens and environment were, and still are, full of medicinal and edible plants. On a larger scale, humans are losing their ability to appreciate how important plants are to the biosphere and to our lives. When orchids flower, however, they emerge from the green blur, reflecting other colours of the rainbow back to our eyes, forcing us to take note. Orchids are the 'vitamin A' we all need to allow us to see again.

Early exposure to the natural world is key to shaping a life-long interest in nature. The world was a very different place just 70 years ago. Whereas many older people mourn the decline in the numbers of birds they see today compared to their youth, a younger generation is largely oblivious to a past brimming with life. Each new generation views the environment into which it is born as being somehow 'normal'. This age-related difference in perceptions between generations is referred to as 'generational amnesia' or 'Shifting Baseline Syndrome'. We are all teachers and storytellers, and it is incumbent on the older generation to record, describe and explain what life was like in the recent past. But, if we wish future generations to become more engaged with nature, and able to enjoy the diversity that we see today, it is important to pass on our knowledge to young people and, above all, to inspire them.

Cattleya mossiae

'Large populations of Cattleya mossiae *still remain in the wild, especially in some parts of the Andes Mountains. But deforestation and predation of this species, even inside protected areas, puts this species in great risk of extinction. In a few decades* C. mossiae *could be as scarce as* C. lueddemanniana *and* C. gaskelliana *are today.*' Castiglione, Pahl, and Vaamonde, 2021

When travelling by air over South America, it is not unusual for the passengers to clap and cheer when the aircraft touches down. Whether this is sheer relief at landing safely at their destination, or Latin American exuberance, we have never been sure, but prefer to think it is the latter. One doesn't expect such a reaction at a scientific orchid conference, but a talk by Gerardo Castiglione about strategies for conserving *Cattleya mossiae* in a small Andean village in Venezuela had the audience clapping and cheering at regular intervals. Gerardo was speaking at the Sixth Scientific Conference on Andean Orchids, held in 2019 at EAFIT University in Medellín, Colombia, and his talk about the efforts of a small community to conserve the orchid was truly inspirational.

Endemic to Venezuela, *Cattleya mossiae*, with its beautiful, large, long-lasting, and fragrant blooms, is the country's national flower. It grows in semi-deciduous forest transitioning into rainforest, in a wide area along the Cordillera de la Costa, a mountain range extending along Venezuela's Caribbean coast, on both the north-facing and south-facing slopes 800–1,500 m (2,600–4,900 ft) above sea level. *Cattleya mossiae* can also be found growing on the south-facing slopes of the Venezuelan Andes. Its story is the familiar one of habitat conversion and the systematic looting of plants, to the point where large areas have been completely denuded of this lovely species, especially close to the large towns and cities. Until the 1950s, it was common to see large colonies growing on the trunks and branches of bucares (*Erythrina poeppigiana*), ceiba species, and other trees on the Cerro El Ávila, a mountain on the edge of Caracas. Today the largest remaining populations are found in the states of Portuguesa and Mérida in the Venezuelan Andes.

Gerardo Castiglione described how an educational initiative in Aricagua – a village of around 4,500 people tucked away in the folds of the Venezuelan Andes in the Tapo-Caparo National Park, 1,225 m (4,000 ft) above sea level – inspired the local people to cultivate the orchid in substantial numbers. As Castiglione said, the key to ensuring a plant's survival is to teach the people to love and appreciate it.

Time and again, when talking to people who have successfully established reserves, they stress the importance of involving local communities. Teaching them about the global importance of their natural heritage instils a sense of pride. Most of the projects we have highlighted have a strong educational focus, involving people of all ages and all walks of life: growers, farmers, public officials, and in some countries, the police, and the military. The aim is to protect both natural and cultural heritage. Local acceptance requires coordination with the appropriate authorities, an understanding of local politics, local leaders, and schoolteachers. Educational campaigns through workshops in schools and community centres stimulate engagement and raise environmental awareness among both children and adults, showing them the importance of caring for their local ecosystems.

Darling, South Africa
The hot, dry summers and cool, moist winters experienced by plants in the Mediterranean are also experienced in Eastern and Western Australia, California, the Cape Region of South Africa, and Chile. These areas have one thing in common – they are orchid hotspots. South Africa has close to 500 orchid species, mostly terrestrial, with new discoveries still being made each year. Its most famous species is *Disa uniflora*, which grows on Table Mountain. The Cape Floristic Province in South Africa is one of the world's major centres of biodiversity for temperate flora, with around 9,000 plant species in an area of less than 90,000 square kilometres (35,000 square miles). Included are 241 orchids, of which 68% are endemic to the region. This incredible diversity of plants reflects the diversity of landscapes, principally the fynbos biome. Threatened by agriculture, urban development, and encroachment by invasive non-native trees such as pines, fynbos is a botanist's paradise.

   In September each year, the Darling Wildflower Show shares the Darling region's botanical riches with thousands of visitors. Landscaped wildflower displays are constructed that represent a journey through the varied natural habitats that occur in the region: endangered renosterveld (low-lying inland and granite koppies), sandveld (low-lying inland fynbos moving towards the coast), strandveld (low-lying coastal fynbos), and wetland areas. Specimens are collected by volunteers with permits from CapeNature, and with the permission of landowners and local farmers. Volunteers adhere to strict rules and collect only a few specimens from each population, leaving the rest to produce seeds in the following years. The specimen table displays are particularly popular with visitors eager to learn the names of the plants making up their own local flora. Examples of local flowers (including orchids), each accompanied by its local and botanical name, are arranged in small specimen vases.

Gardens by the Bay
Nothing can prepare you for the exhilarating experience of stepping from the hot and humid outdoors into the cool of the

Cloud forest at Singapore's Gardens by the Bay.
Photo: Philip Seaton

191　EDUCATION: THE TEACHERS AND THE STORYTELLERS

cloud forest conservatory at Singapore's Gardens by the Bay. As you stand in front of a waterfall cascading 35 m (115 ft) from three quarters of the way up the side of the 42 m (138 ft) 'cloud mountain', the sheer scale of the surroundings takes your breath away. Take the elevator to the top if you must, but it is better to meander slowly up the gently sloping ramp that spirals upwards through the cool, moist conditions created by the plants and the mists hissing from humidifiers. The 'cloud mountain' is an intricate structure completely clad in orchids, ferns, peacock ferns, clubmosses, bromeliads, and anthuriums. Interpretation boards explain that orchids grow at different elevations in tropical mountain regions 1,000–3,000 m (3,300–9,800 ft) above sea level, found in South-East Asia, and Central and South America. The majority of orchids are tiny, and to enable visitors to appreciate their intricate beauty, they can be viewed through magnifying glasses.

VIEWPOINT

Orchids and education in Chile's Mediterranean region
by Sergio Elórtegui Francioli
(professional photographer and teacher)

'¡No señor, en Chile no hay orquídeas!' ('No sir, there are no orchids in Chile!')

As in many temperate countries around the world, most people in Chile think that orchids only exist far away in the tropics, and are unaware of the orchids that grow naturally on their doorsteps.

Nothing can replace getting out into the field. It is much more exciting to see for yourself how an orchid is able to glue its pollinia to the back of the pollinating insect than to read about it in books. Such an experience will never be forgotten. Seeing how plants and animals live and interact brings a much deeper understanding of the living world, and can lead to the beginnings of a scientific approach in the minds of children. Orchids are beautiful and strange. They are long-lived and wait for us year after year in the same place. They have a complex anatomy and development, and their distinctive forms are unmistakeable. Even the most restless children become interested when we talk about 'sexual parasitism' and how some orchids use the mating desires of male bees to reproduce – how the insects eventually die exhausted without receiving so much as a drop of nectar in return!

Chile is a biodiversity hotspot, with around 72 orchid species, all terrestrials. Of these, more than 50% are concentrated in the central Mediterranean region, all of them endemic. Paradoxically, many thrive in the same places as humans: coasts, dunes, valleys and woodlands. Many are small – practically invisible – and, due to farming practices and uncontrolled urban expansion, consist of just a few individuals. About 90% of the sclerophyllous forests (those adapted to drought and heat) have already been degraded or have disappeared. A recent survey of orchids along the coast revealed that the greatest number were found around the edges of towns and plantations. Many of these populations have since disappeared, and orchids are no longer found within towns and cities.

Students at the Colegio Sagrada Familia, Reñaca, in Viña del Mar, Valparaíso, have been taken into the field to study a population of *Bipinnula fimbriata* in las Dunas de Concón, gathering data about growth, flowering, and fruiting. This helps them to understand the orchid's life cycle, and to formulate their own scientific questions. What is the pollinator?

When does pollination happen? Why do the flowers have 'beards'? Students learn to value their own heritage and to ask questions on much broader issues, such as, 'Why are our towns increasingly taking over natural spaces?'

Over the years, many students have become interested in studying their local orchids in detail. Some of their projects have been published in scientific periodicals and have been presented at botanical congresses. The information has been incorporated into what little is known about Chilean orchids. Today there are various co-operative projects in progress involving 'children and scientists' in collaboration with staff at the botanical laboratory of the Catholic University of Valparaíso and at the Botanic Garden of Viña del Mar. For example, they measure the effect of inbreeding on fruit set, and the effect of crossing separate populations on germination. Students are investigating the distribution of orchids, registering any new locations at the herbarium of the National Museum of Natural History. Through their direct contact with orchids, the children are learning to ask questions, and develop a lifelong interest in their environment. Never again will they fail to 'see' these plants.

Working with children has brought together many friends and institutions, both in Chile and abroad. Ecology workshops that bring together children from various schools in central Chile take place under the umbrella of different projects hosted by different institutions, or workshops occur spontaneously, according to each school's interests. Today, central Chile is the focus of multiple sociological crises that translate into the loss of biodiversity, which has been exacerbated by one of the longest droughts in the country's history (15 years). Under these conditions, many schools are becoming increasingly keen to tackle environmental issues.

For many years the focus of Chilean botanists has been on searching for new orchid populations and species that have not been seen in recent decades. Many of the investigations centred on learning the orchids' life cycles and their habitat requirements. Soil samples harbouring mycorrhizal fungi have been taken to elucidate the orchids' needs and obtain a better understanding of the reasons underlying their irregular distribution. I believe that the most important outcome is when the school becomes a centre for socio-environmental activity in its local community. The school contributes its creativity, assumes responsibility through talks and events, and shares the natural values with its neighbours and the regional authorities. In this context, orchids, beautiful and unknown, can become 'icons' or 'flagships' in the fight for the conservation of local ecosystems.

The 'orchid of Valparaíso' (*Chloraea disoides*) is a species for which no more than a hundred individuals are known in the region. Today it is the subject of a strong publicity campaign, initially orchestrated by children, to save the last remaining populations. Meanwhile, many populations continue to be destroyed by uncontrolled urban expansion on the edges of towns and cities. The only option for some of these orchids is to transplant them from the natural site to a new area. Otherwise they would perish. Together with the National Botanic Garden and two foundations, it has been possible to transplant some complete populations to new areas that share similar characteristics and are better protected. Thus far, these efforts have been very successful. Some schools have modified parts of their gardens and converted them into ecosystems where the local orchids can grow. Colegio Sagrada Familia, for example, successfully maintains a population of 50 established plants of *Bipinnula fimbriata*, an orchid of the dunes. After 15 years, they now have a hundred small plants that have germinated under the shade of the transplanted plants.

It has been a difficult but thrilling journey. Every year when they flower one feels the orchids' affection like a loving pat on the back. Currently, the main focus has been on 14 rescued plants of critically endangered *Chloraea disoides*. Its pollinator remains unknown and is a subject of further

(left)
Construction of a school herbarium in Chile.
Photo: Sergio Elórtegui Francioli

(opposite)
Rescuing *Bipinnula fimbriata* from urban expansion.
Photo: Sergio Elórtegui Francioli

research. Some students are investigating its survival following transplantation, and its reproduction following outcrossing with other populations with few surviving individuals, separated by hundreds of kilometres due to habitat fragmentation. After five years, these experiments have already yielded positive germination results. Moreover, we have found that the work of collecting pollinia and cross-pollinating flowers by hand is best achieved by the small fingers of groups of young people committed to saving the species they study. Children can bring about our much hoped for change for the better. Let's take them outside, put them on Nature's stage, and talk seriously and passionately about conserving their natural heritage!

195 EDUCATION: THE TEACHERS AND THE STORYTELLERS

Guadalupe

Travelling down the valley of the Río Magdalena south from Neiva, in the District of Huila, Colombia, each small town or village has *Cattleya trianae* growing in the trees in its central plaza. Visitors eventually arrive in the municipality of Guadalupe, in a small valley on the margin of the Río Suaza, on the flanks of the Eastern Cordillera. In 1980, a decree was signed ordering the planting of trees on either side of the road for one kilometre before each town. Entering Guadalupe, drivers pass through a tunnel of trees full of *C. trianae*. The orchids have been planted by children from a school at the end of the road, the Colegio María Auxiliadora, under the supervision of their teacher Alvaro Tobar. The trees are planted with many different coloured varieties – ranging from pale lilac to dark purple. Some have contrasting deep magenta lips. This makes a truly magnificent sight when all of the cattleyas are in full bloom. Signs nailed to some of the trees invite people to care for their plants. One says: '*Cuidemos los arboles ¡Son fuente de vida!*' – 'Look after the trees, they are the fountain of life!'

Sometimes you just want to say, 'Thank you!' Thank you to the people who had the vision to order the planting of the trees. Thank you to the teacher with the vision to plant the trees with orchids. Both will be enjoyed by generations to come.

In the lab

Students love putting on a white coat. They feel like scientists (and indeed they are). Working in a laboratory gives them an opportunity to make a positive contribution to orchid conservation, while at the same time learning more about plant biology. What could be more exciting than growing a plant from a tiny seed? Looking down a microscope reveals a new world. Students can observe the seeds of tropical orchids germinate and develop into tiny green protocorms on different nutrient media, or watch the slender branching threads of mycorrhizal fungi fan out across a porridge-based medium until they infect the seeds of temperate and tropical species. As part of a programme of extra curricular activities, 16- and 17-year-old students at King Charles I School in Kidderminster in the UK, for example, have been growing British native orchids from seed for reintroduction in the local area, and gathering new data on the lifespan of orchid seeds when stored at refrigerator temperatures.

In the UK, perhaps the best-known orchid-related school enterprise is the Writhlington School Orchid Project, founded by teacher Simon Pugh-Jones in 1991. Six glasshouses, housing more than 800 orchid species, have been managed by teams of students, some of whom have had the opportunity to become involved in orchid conservation initiatives around the world, working with schools, communities, and conservation groups in Africa, Asia, and the Americas. Students have set up orchid propagation laboratories

A school in Guadalupe where young children learn about their native orchids.
Photo: Philip Seaton

in Rwanda and Laos, for example. Student teams have also led a number of orchid education initiatives to develop public displays of plants both in the UK and abroad.

In the USA, the Million Orchid Project, established at Fairchild Tropical Botanic Garden in Florida, is the largest educational outreach programme dedicated to orchid conservation in the country. The project's goal is to propagate and establish one million native orchids (mostly epiphytes) in urban landscapes throughout the region. Led by Jason Downing, over half a million orchids have been planted out as of 2022, and are currently being monitored with the help of schoolchildren and volunteers from the local community. Over 20 community partners and more than one hundred schools grow nine rare orchid species in classrooms and surrounding neighbourhoods. A decommissioned school bus, now fitted out as a mobile micropropagation lab by students from the University of Miami School of Architecture, is deployed to bring the lab to the students. With its livery displaying the Florida ghost orchid, the bus bridges the gap between many different schools, and generates public awareness while on the road between visits. As an ongoing experiment in itself, the Million Orchid Project continues to make scientific discoveries that involve orchid establishment and survival while, at the same time, educating and informing the public under the framework of citizen science.

Participants in the Scientific Orchid Conference on Andean Orchids, held in August 2019 at EAFIT University in the heart of Medellín, Colombia, were delighted to find that trees on the university campus had been planted with orchids. They were

The 'mobile orchid lab' used by Fairchild Tropical Botanic Garden to drive between schools in the Miami area for the Million Orchid Project.
Photo courtesy of Jason Downing

clearly thriving, as evidenced by their extensive root systems clinging to the trunks of the trees. Many were in bloom. The enrichment of the university campus began with a donation of 20,000 cattleyas by Mario Posada Ochoa. Three quarters of the plants were hybrids, and the remainder were from Santander, Chocó, Valle de Cauca, and the Llanos Orientales. Among their South American cousins were a sprinkling of 'exotic' species, including the Asian *Vanda tricolor* var. *suavis*.

In 2017, scientists and technicians at the university's micropropagation lab began developing protocols for growing native orchids as part of their conservation strategy. Their germplasm bank contains seeds of many of the regionally endangered orchids, including *Cattleya warscewiczii, C. dowiana* and *Miltoniopsis vexillaria*. A programme of regular testing has been implemented, including germinating seed on agar-based media and staining the seeds with 1% tetrazolium. Staining allows the researchers to monitor the percentage germination of a range of different species over time, and to identify any that may have especially short-lived seeds. To date, seeds have all been successfully germinated using asymbiotic techniques. In the long term, the aim is to employ symbiotic techniques for those orchids being raised for forest restoration projects.

In addition to the laboratory, there is a greenhouse large enough to accommodate around 10,000 plants. At the time of writing, around 7,000 orchids have been acclimatised there. Novel mixtures of nutrients prepared in the laboratory are being evaluated in the greenhouse, often resulting in improved growth and development. To continue and expand the university's conservation activities, there is always the need to attract new streams of funding. A small income is generated through the sale of a range of species, both as small seedlings in flasks, or as larger seedlings that have been acclimatised within the greenhouse. They are sold together with a small booklet of instructions for after-care. The laboratory staff offer propagation courses to both commercial and amateur growers. The wider aim is to teach people to appreciate the natural environment and to value their local biodiversity, and to engage with other partners such as the Parque Arví nature reserve, in reintroduction and restoration projects.

The best of both worlds
Few people have a foot planted in both worlds – one foot in the world of horticulture, and one foot in the scientific world. Orchid growers are seldom seen at scientific meetings, and scientists seldom visit orchid shows. Nelson Machado-Neto is an exception. Fascinated by their colours, scents, and forms, he has been able to combine what began as a hobby with his professional life at the University of Western São Paulo (UNOESTE), in the municipality of Presidente Prudente, Brazil.

    Machado-Neto grows around 5,000 plants in two shade houses, 2,000 of which are true species. Most are grouped in small numbers of two, three or four plants; for others, there are multiple entries consisting of 10 to 20 plants, especially the lithophytes, which are his passion. There is a large population of *Cattleya brevicaulis* of more than 200 plants, and around 50 plants each of *C. purpurata*, *C. labiata*, *C. walkeriana*, and *C. tigrina*. He only buys plants from reputable nurseries, with as many species as possible purchased as seedlings in flasks. In the latter case, Machado-Neto can then be certain that those plants were not harvested from their natural habitats. Collection of seed from the wild, either by himself or a trusted colleague, has only been carried out after the appropriate permissions have been secured. Only a small number of capsules are collected at any one time to ensure that some plants can spawn seedlings naturally.

    Presidente Prudente is on the border between the Brazilian Savannah and the Inner Atlantic Rain Forest (semi-deciduous forest), with many riparian forests. The climate is tropical, with a relatively cool and dry season from June to August, and a hot – 35°C (95°F) or more – rainy season from November to March. In 2021 there were three light frosts. In recent years, the climate has become hotter and drier, making growing orchids more challenging. Machado-Neto has only been able to grow his collection of rock-growing cattleyas because he uses sprinklers, located one metre above the plants on the benches, to maintain moist/humid micro-environments. Because many of his plants were laboratory-grown, and either bought as seedlings or raised in his laboratory, they have been subject to Darwinian selection pressure – those that survive in the warmer conditions are propagated, whereas those that cannot tolerate greenhouse/shade house conditions perish. This raises an interesting question. Would it be possible, or desirable, to raise plants in living collections that were better adapted to the forthcoming increases in environmental temperature for future reintroductions?

Machado-Neto is specifically interested in developing techniques to store and germinate orchid seeds (see 'Orchid Seed Stores for Sustainable Use' (OSSSU), page 276) without the need for expensive equipment. His wife, Ceci Castilho Custódio, has been working on seed storage and conservation alongside him at UNOESTE since 1996. Between them, they have trained 92 students during the 15 years since they established a UNOESTE orchid seed bank. Machado-Neto uses his private collection to teach students how to grow the plants, pollinate the flowers, and harvest and store the seed. Storage is the main focus of the group, and they are implementing other seed conservation techniques, including cryopreservation. His research group continues to grow in number, with many students continuing to make important contributions to understanding the problems inherent in the long-term storage of orchid seeds. Initially working mainly with *Cattleya* species, the work has recently expanded to include: the reintroduction of tropical terrestrials; tissue culture of members of the Oncidiinae, *Catasetum* and *Vanilla*; and seed germination. Soon they will receive seeds from Luciano Zandoná's project (see 'Cantareira', page 172).

Orchids and artists
The beauty of orchids has attracted botanical illustrators and other artists down the ages. Two such individuals are Ian Cartwright and Ruth Grant. They both studied natural history illustration at Blackpool and The Fylde College (University of Lancaster) in the UK, producing colourful posters that promote orchid conservation.

Cartwright's poster illustrates the relationship between the bucket orchid, *Coryanthes kaiseriana*, and its associated bees and ants, against a background of the orchid's South American habitat, characterised by trees cloaked in bromeliads and other epiphytes. The plant is rooted in an 'ant garden', essentially a colony or nest of numerous *Camponotus* or *Azteca* ants. The flower structure is bizarre. A cup-shaped lip contains a watery fluid produced by glands at the base of the column. Male euglossine bees are attracted by the orchid's perfume and, upon landing on the flower, they slip and fall into the 'bucket'. Smaller bees cannot fly if their wings are wet, and there is no room in the lip for larger bees to spread their wings. The inner surface of the lip is smooth and slippery, but there is an escape route – a virtual stairway leading out beneath the column. In crawling through this opening, the bee first passes beneath the stigma and then the anther, where the pollinia are firmly glued between its thorax and abdomen. The bee flies off and, not having learned its lesson, repeats the process on either the same, or another flower. This time the pollen is removed and deposited on the flower's stigma as the bee makes its bid for freedom.

Grant was funded by the Royal Horticultural Society to visit Mount Kinabalu in the East Malaysian state of Sabah, which

Nelson Machado-Neto, Ceci Castilho Custódio and students with a population of *Cattleya brevicaulis* growing in Nelson's orchid shade house in Brazil.
Photo: Nelson Machado-Neto

(overleaf left hand page)
Poster illustrating *Paphiopedilum rothschildianum*, with *Paphiopedilum dayanum* in the background, growing on Mount Kinabalu.
Painting by Ruth Grant

(overleaf right hand page)
Poster illustrating *Coryanthes kaiseriana* with associated ants, and the orchid's euglossine bee pollinator.
Painting by Ian Cartwright

allowed her to visit the habitat of Rothschild's slipper orchid (*Paphiopedilum rothschildianum*). Rising to 4,100 m (1,345 ft), Mount Kinabalu is a dream destination for orchid-lovers. It is home to one of the world's richest floras, with an estimated 6,000 plant species, including more than 830 orchids. Here, the highest density of orchids is found in the montane forest at around 500–2,000 m (1,640–6,560 ft) elevation. *Paphiopedilum rothschildianum* has been found in only three places on the mountain. In two of these, it has either been picked over by collectors, or destroyed. Growing with it is an equally rare slipper orchid, *P. dayanum*, distinguished by its mottled foliage and smaller flowers with unspotted purple petals that have distinctive eyelashes on their margins. Grant's illustration depicts both orchid species, with the mountain in the background. She painted the picture after being taken to a spot where the orchid once occurred, and illustrated the plants from cultivated specimens. The two locations where it survives remain a closely guarded secret, as it continues to be targeted by poachers.

ORCHID CONSERVATION INTERNATIONAL
Saving The World's Orchids

MOUNT KINABALU AND ITS UNIQUE ORCHID FLORA

Mount Kinabalu, in the East Malaysian state of Sabah, rises to 4101 m, dwarfing the surrounding hills and mountains. It is home to one of the richest floras on earth with an estimated 6,000 species so far recorded. The richness of the flora can be traced to the many different habitats found on the mountain, including extensive areas of ultramafic (serpentine) soils which have a very peculiar flora. The richest area for orchids is the montane forest between about 500 and 2000m elevation. Orchids grow there on the ground, on rocks and on trees, from their trunks to their smallest twigs. Over 800 species of orchids have so far been recorded on Kinabalu. They grow almost everywhere on the mountain from its base to the summit plateau.

The most famous orchid on the mountain is Rothschild's slipper orchid (*Paphiopedilum rothschildianum*), discovered in 1887 by a zoological collector and thought to be extinct until rediscovered in 1960 when orchids were being collected to decorate the pavilion for HRH Prince Philip's visit to Sabah. It grows on ultramafic soils on narrow ledges of steep cliffs under light forest cover. It has been found only in three places on the mountain, in two of which it has been either collected out or destroyed. Growing with it is the equally rare *Paphiopedilum dayanum*, distinguished by its mottled foliage and smaller flowers with unspotted purple petals that have distinctive eyelashes on their margins. *P. dayanum* is known from only two localities on the mountain, growing under bamboo in the leaf litter of the forest floor.

Many orchids are threatened by forest clearance. Kinabalu's status as a National Park provides some protection but showy orchids are still collected for trade within the park's boundaries.
Orchid Conservation International supports orchid conservation projects worldwide.

Orchid Conservation International is a Registered Charity (No.1147762)

BLACKPOOL AND THE FYLDE COLLEGE
An Associate College of Lancaster University

info@orchidconservation.org

ORCHID CONSERVATION INTERNATIONAL
SAVING THE WORLD'S ORCHIDS

CORYANTHES
TROPICAL AMERICAN BUCKET ORCHID

This is the Tropical American Bucket Orchid (Coryanthes Kaiseriana) one of 34 species within the genus Coryanthes, it is found in Mexico, Guatemala, Honduras, Costa Rica, Panama, Trinidad, Venezuela, British Guiana, Brazil, Colombia and Peru. The orchids grow as epiphytes in low lying hot and humid regions. They are particularly adapted to the highly acidic conditions found in the nests of Camponotus and Azteca ants. The flower is pollinated by the Euglossine bee, a noisy and aggressive insect. The bees fall into the bucket, which is filled with a sticky fluid secreted from a mucilaginous gland on the flower. This enables the pollen to attach itself to the bee's back when it escapes through a side passage in the bucket.

info@orchidconservation.org

EDUCATION: THE TEACHERS AND THE STORYTELLERS

14 Reserves

In October 2021, the UN Global Biodiversity Framework proposed 21 targets for 2030, which include protecting at least 30% of the world's land and oceans. The gold standard of orchid conservation is the preservation of orchid-rich habitats. Many have already disappeared, and there is an urgent need to preserve the surviving remnants and, where possible, to acquire additional land for habitat restoration. How is this to be achieved? National governments have an important role to play in establishing national parks, for example, but it is often NGOs (non-governmental organisations) that drive local conservation initiatives. What is the purpose of reserves? To capture the local biodiversity, certainly, but more than this it is research and education. Involving the local community instils a sense of ownership, and of pride in preserving these precious natural resources.

A network of Andean orchid reserves
The Tropical Andes stands unequalled among the world's biodiversity hotspots as measured by species richness and endemism. It contains about one-sixth of the world's plant life, making it the top hotspot for plant diversity. It also harbours the largest variety of amphibians, birds, and mammals. For reptiles, it ranks second after the Mesoamerica hotspot. Peru, Ecuador, Colombia, and Venezuela, with their plethora of microhabitats, have seen an explosive speciation of orchids. They are the hottest of hotspots for orchid biodiversity. Colombia, for example, has to date recorded 4,270 named species of orchids, with a similar number in a much smaller area in neighbouring Ecuador. In comparison, Brazil has 2,962 recorded species, but is a vastly larger country.

Many orchid species occupy specific habitats that occur in very small areas, and a network of reserves is required to safeguard this diversity. Often the existing private reserves in the Tropical Andes are small, occupying only a few hundred hectares or less. Where they have become isolated, essentially forming an archipelago of small islands in a 'sea of agriculture', over time they will inevitably lose species. One of the aims of the EcoMinga Foundation and Corporación SalvaMontes is to connect reserves with wildlife corridors through land purchase and agreements with local landowners. Where they connect areas of the mountainside at different elevations, corridors allow the seasonal migration of birds and butterflies and enable the upward migration of plant species. As the climate changes, such corridors may be crucial for the survival of many species by allowing migration to higher (cooler) elevations – at least for animals. Whether the current pace of climate change will allow the natural migration of plants to take place remains an open question. Orchids may have an edge because their seeds are dispersed by wind, but because some species may only grow on specific host trees, orchids will likely perish unless

(previous pages)
Cloud forest in the Colombian Andes.
Photo: Philip Seaton

suitable trees are already present at higher elevations. The seeds of some host trees are dispersed by animals, and if animals are provided with a corridor for upward migration, orchids may eventually follow. Just as the tops of mountains served as climatic 'refugia' during past ice ages, they may perform a similar function on a warming planet.

A word about trees
'It is reasonable to assume there was continuity of epiphytes, particularly orchid species, with only minor endemic exceptions. Back then, the humid mists in the underforest must have been the same on both sides of the watershed. The rivers would have been permanently full (unlike today). The dense garden of epiphytic bromeliads created a huge suspended lake because of the water kept in their rosettes, functioning as a regulator of humidity and temperature during dry periods.' Miller, Warren, Miller, and Seehawer, describing Brazil's Atlantic Rainforest, 2008

Most tropical orchids need trees. Trees are the scaffolding that supports a rich epiphyte flora of lichens, mosses, ferns, and fungi that form a bed for germinating orchid seeds. Orchids often co-exist with other groups of flowering plants that are adapted to life in the canopy, such as aroids, pipers, and in the New World, bromeliads and even cacti. Old, mature, trees are particularly important because they support many different orchid species during the course of their lifetimes. A single tree may, for example, be colonised by more than 50 different orchid species. Meanwhile, terrestrial orchids live in its shade, rooted in the ground below. Thus, if we wish to conserve orchids, we must also conserve trees and plant more of them. But which trees? A hectare of tropical cloud forest will contain many different species, but not all trees make equally good orchid hosts. Some orchids can be very choosy indeed, growing on just a handful of favoured tree species. *Barkeria* species, for instance, often show distinct host specificity, favouring trees with a spongy or papery bark that retains moisture during long dry periods. *Lepanthes tibouchinicola*, which is endemic to Antioquia, Colombia, grows almost exclusively on *Andesanthus lepidotus*. The endemic *L. caritensis* occurs solely in Carite State Forest in Puerto Rico, and grows only on a single species of tree (*Micropholis guyanensis*), where it is associated with high moss coverage.

The Colombian Orchid Society has established a tree nursery at their reserve near the small town of Jardín. Seedlings of endangered species such as Colombia's national tree, the wax palm (*Ceroxylon quindiuense*), are cultivated alongside important timber trees that include black cedar (*Juglans neotropica*), oak (*Quercus humboldtii*), and *Cecropia* species. Although the cloud forest reserve is in good condition, many trees that were valuable for their timber were removed in the past, and the species composition of the forest has changed. In addition, there are areas that had been converted into pasture for livestock. With the cattle removed, an extensive tree planting programme is under way to restore the woodland. A small experimental plot is devoted to studying restoration techniques and plant succession – how species composition changes over time. Through the gradual process of succession, a more complex environment will evolve, with mammals and birds helping to disperse the seeds of other trees and shrubs into the newly restored habitat.

Corporación SalvaMontes was established with the aim of conserving three endangered *Magnolia* species (fewer than 35 adult specimens of one of these, *Magnolia polyhypsophylla*, exist in the wild). It has set up three reserves at 'Alto de Ventanas', at an altitude of 900–2,400 m (2,950–7,870 ft) at the northern end of the Central Cordillera in the Andes. Magnolias are suitable orchid hosts, and several orchid species can be found growing on them at 'Alto de Ventanas', including *Miltoniopsis vexillaria, Masdevallia macrura, M. ventricularia, Specklinia colombiana* (syn. *Acostaea colombiana*), *Pleurothallis acinaciformis*, and a number of *Lepanthes* species.

Magnolias are an ancient genus of 'primitive angiosperms' that evolved around 95 million years ago, when dinosaurs roamed the Earth, and when bees, butterflies (and orchids) began to diversify. Their flowers are often pollinated by deliciously named 'tumbling flower beetles' that are attracted by their perfume and the promise of protein-rich pollen. The seeds take around 6–9 months to germinate in a tree nursery, their roots gradually becoming colonised by mycorrhizal fungi, which are different from those that connect to photosynthetic orchids. At Corporación SalvaMontes, once a seedling reaches four years of age and is about 1 metre in height, it is transplanted to one of the reserves. Over time, its trunk and branches become home to a succession of epiphytes: lichens, mosses, ferns, bromeliads. At around ten years of age, or perhaps a little younger, some of the thousands of orchid seeds will encounter a compatible fungus and germinate high in the tree's canopy or on its trunk. The tree will probably begin blooming at around 15 to 20 years of age. It may live for another 200–300 years, all the while fixing carbon dioxide, thereby reducing the effects of global climate change.

Cloud forests play a vital role in the provision of freshwater, helping to preserve watersheds and other 'ecosystem services'. The Colombian Orchid Society reserve, for example, is the main water source for coffee fincas, for people who live downstream, and for forests at lower elevations, such as xerophytic (dry) forests. Cloud forests act as sponges, and release the water gradually rather than as a deluge, thereby playing a crucial role in averting flooding. In the 1700s, forests in the future park around Río de Janeiro were cleared for fuel, coffee growing, and livestock. The small streams in the former forest had been a significant source of the city's water supply, and with variable rainfall, the city began to experience water shortages and flash floods. Today, the surrounding forests are largely the result of reforestation. The importance of the water they provide should not be underestimated or taken for granted.

VIEWPOINT

A personal story from Colombia
by Ana María Sánchez-Cuervo

I was lucky to have been born in Colombia, one of the world's most biodiverse countries. I was even luckier to have been born in the tropical Andes, an area that has been identified as a global biodiversity hotspot. My home town is Tunja, the capital of the Department of Boyacá, and Colombia's highest capital city. I have always been fascinated with my mother's stories about the unique habitats and the diversity of animals that she saw as a child in the 1950s. She talked about our extensive *páramos* (alpine tundra), cloud forests, wetlands, small mammals, and all kinds of birds and plants. My mom told me she used to hike up to the high *páramos* to look for native orchids to eat, as their pseudobulbs were a good source of food and water. Over time, she took every opportunity to get my brothers and I in touch with nature around Tunja, with the hope that we could have the same experiences she had in the past. I remember seeing rhinoceros beetles the size of a baseball, thousands of May beetles, and bats. Unfortunately, I also remember how it was much harder to find all of the birds and plants that my mom had described to us, especially the orchids.

I am painfully aware that all that remains for my young nieces are our memories and stories about the incredible biodiversity we used to have in the high Andean ecosystems. They always wonder why there is nothing left for them to see and enjoy. The *páramos*, the cloud forests, and the vast wetlands have been reduced to small fragments, or they have disappeared altogether in many areas along with our native fauna and flora. Deforestation and fragmentation are mainly driven by agricultural practices, livestock activity, and firewood harvesting, as these are the main source of income and energy for most of the local population. Fortunately, there are multiple institutions and initiatives working hard to recover, reconnect, and protect these natural areas as well as to educate the next generation in the Andean highlands. Because of their efforts, I am hopeful that my mother's great-grandchildren will be able to experience the incredible nature in which she grew up. But I also know that we have a long way to go before we can accomplish that dream.

Ecotourism
Not everyone who encounters an orchid has the urge to dig it up or pluck it from its tree host. Most people get satisfaction simply by seeing the orchid growing in its natural habitat, where it lives in a world rich in natural wonders. This is the basis of ecotourism. Some wish to photograph their orchid 'prey' – although

photographers should be wary of posting the coordinates of rare and endangered species, which poachers can use. Likewise, people should be aware that when photographing, or simply looking at orchids, especially terrestrials, they need to avoid inadvertently treading on seedlings. Trampling the soil can also lead to compaction, which has an adverse effect on the long-term survival of plants. With reasonable care, however, ecotourism provides a valuable source of funding, and provides an economic incentive for local people to preserve wild populations. Its success, however, depends on a steady influx of wealthy tourists, and is vulnerable to unforeseen global events, as exemplified by the recent COVID pandemic.

Orchid tourism must be practised responsibly given the potentially large numbers of visitors. More than 25,000 people climb Mt Kinabalu in Borneo each year, for example, though of course not all of them have come to see the orchids. Professional tour guides have their hands full in pointing out many kinds of life forms, but they also must discourage their charges from collecting plants. Because of its remoteness, balancing an influx of visitors against the welfare of delicate ecosystems is not a problem in Madagascar, however. Tucked away in the Southern Hemisphere far from Europe and Australia, it remains a distant and expensive outpost for many ecotourists who might want to experience its unique biological diversity. Consequently, ecotourism will probably never provide the island nation's unique life forms with the security blanket they so desperately need, because there is little economic incentive to conserve their habitats.

Alto de Ventanas

The Department of Antioquia in Colombia's northern Andes is the country's hottest of hotspots for orchid diversity. Little of the native forest remains, however, as most of the original forest was cut down during the second half of the 20th century. The remaining biodiversity is confined to remnants embedded in a greatly transformed countryside.

Dracula lemurella was first discovered in 1974, in Quebrada El Oro, growing at intermediate elevations, around 1,650–1,800 m (5,400–5,900 ft). A description of the type specimen was published in 1981. With its pale, ghost-like 'face', *D. lemurella* is popularly known as '*mono fantasma*', the ghost monkey orchid. It is only known from a single type locality, but the quality of its original habitat is degrading and its extent reduced. In the past, the ghost monkey orchid was probably subject to illegal poaching. Exploration of the type locality of Quebrada El Oro in 2011 and 2016, revealed that, although some suitable habitat remains, it is highly fragmented and separated by a matrix of pasture and crops. The researchers failed to find any existing plants. A significant proportion of the forest had been cut down, and in nearby areas

Dracula lemurella with fruit flies (*Drosophila* species). The anther cap of the orchid can be clearly seen attached to the thorax of the fruit fly in the centre of the image.
Photo: Sebastian Vieira

most of the original forest had been replaced by pasture (for cows) and agriculture. There is a need for more research around Quebrada El Oro and its surrounding suitable habitat, but in all likelihood, *D. lemurella* is now critically endangered.

Reserva Natural La Selva de Ventanas has been created to protect the habitat for both *Dracula lemurella* and other endangered plants. A new, small population of *D. lemurella* has recently been found at Reserva Natural Los Magnolios. This reserve is important because, in addition to hosting the only known natural population of *D. lemurella* (around 70 individuals), it contains many other endemic species adapted to this elevation. What is the orchid's preferred habitat? It is currently found in well-preserved forest. Would it be equally happy in secondary growth forest, for example? Possibly not. Researchers are currently carrying out a long-term population study, labelling individual plants and following their progress. Another important question is, what is its pollinator (or pollinators)? The flowers have a mushroom-like scent and mimic the fruiting bodies of fungi. Like other draculas, are they pollinated by fruit flies? Almost certainly. What is the orchid's life cycle? How long does it take to grow from seed to flowering plant? All of the plants sold today appear to be clonal divisions of the original collections from the type locality which were distributed to selected orchid nurseries. Current attempts to propagate the species from seed have had limited success, and there is a need for more research into its preferred growing media.

Colombian Orchid Society

In 2016, the Colombian Orchid Society purchased 200 hectares (500 acres) of cloud forest in the municipality of Jardín. As the opportunity arises, they are continuing to buy adjacent plots of

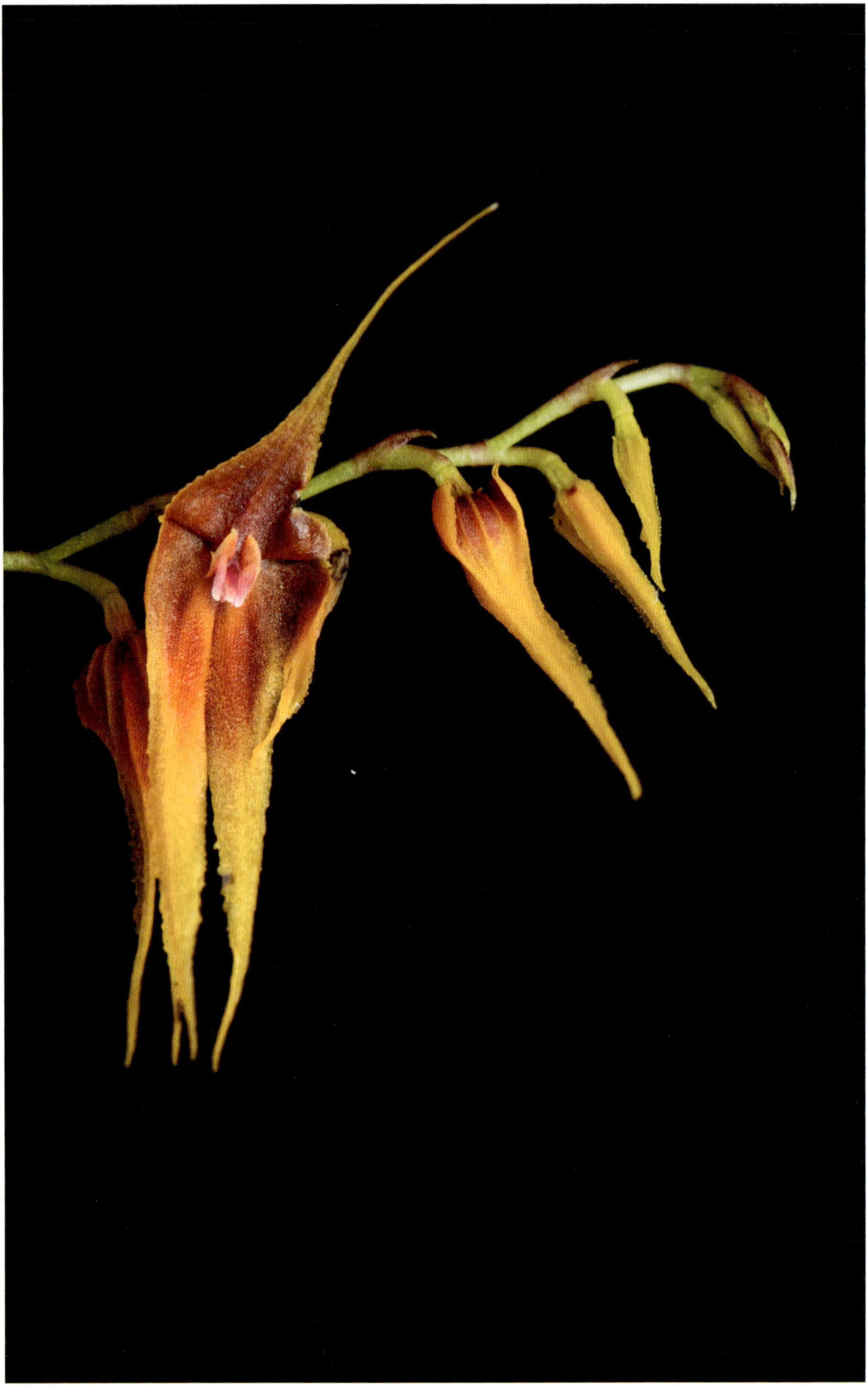

pasture, and, little by little, restoring the forest. Climbing from 2,470 to 3,170 m (9,000 to 10,400 ft) above sea level, it is an area rich in epiphytes. A trail winds its way up the mountainside, punctuated with viewing platforms that provide visitors with a welcome excuse to catch their breath and enjoy the panoramic views. Along the way, a pergola showcases orchid species in bloom. Many orchids, bromeliads, and other attendant epiphytes end up on the ground as a result of branches falling. These plants, which otherwise would have been doomed to die, are rescued and transferred to a large shade house. Here they are grown in pots on open mesh staging, where their phenology (their development over time) can be studied and, once in flower, they can be identified and may be pollinated to provide seed. Alternatively, they can be transplanted along the trail in an appropriate location, or into the forest.

The reserve is rich in terms of both the number of orchid genera and number of orchid species. At the time of writing, more than 146 species had been discovered in 27 genera. Doubtless many others remain to be discovered and, so far, the researchers haven't penetrated the forest as far as the crests of the mountains. Who knows what treasures may be waiting for them? Data are being gathered to determine which trees make the best hosts for orchids, and to pinpoint the location on each tree where each individual species thrives.

The aim is to reintroduce orchid species that were originally present in the reserve but are no longer there – often having been targeted by illegal collectors in the past – such as acinetas, anguloas, and *Miltoniopsis vexillaria*. Members of the Colombian Orchid Society with access to laboratory facilities are using *in vitro* techniques to grow numerous species, with an emphasis on those of horticultural value that have been removed in large numbers, including *Odontoglossum mirandum, O. sceptrum, Masdevallia amanda* and *M. hortensis*, plus those that are naturally scarce, such as telipogons and *Pterostemma antioquiense*.

El Pahuma
The El Pahuma Orchid Reserve in Ecuador was established by the Ceiba Foundation to conserve 375 hectares (925 acres) of tropical mid-montane forest. The young and energetic can spend the two days required to climb up to *El Sendero de los Yumbos* (the Yumbo Trail), where they are rewarded with spectacular views over the Andes. This pre-Inca trail, carved deep into the crest of the mountains, was formed over hundreds of years by the feet of thousands of porters journeying through the mountains. Visitors who ascend the mountain become aware of gradual changes in the flora according to the different microhabitats. The reserve supports more than 300 orchid species, and ornithologists are also attracted to the reserve by its rich avian fauna.

Lepanthes tibouchinicola close-up.
Photo: Sebastian Vieira

Cyrtochilum macranthum is native to the montane cloud forests of Colombia, Ecuador and Peru.
Photo: Philip Seaton

Both terrestrial and epiphytic orchids are regularly encountered along the trail-side, each with its own story to tell. Veitch described *Odontoglossum hallii* as being 'the grandest of the Ecuadorian Odontoglots', and goes on to say that 'it was discovered by the gallant officer whose name it bears, Col. Hall, in 1837 . . . in the Quito district.' *Dracula sodiroi*, with its tubular scarlet blooms, by no means resembles your typical *Dracula* species, which are usually studies in muted browns and creams, resembling the undersides of exotic fungi. An orchid 'garden' has been planted at the foot of the mountain containing many orchids salvaged from the reserve, providing an opportunity for casual visitors to enjoy the native flora of El Pahuma.

The reserve generates enough income to be more-or-less self-sufficient, and Ceiba's original goal of providing local people with a sustainable, non-destructive source of income, is being met. The Lima family, who own the forest, have built a small restaurant close to the reserve to further supplement their income.

Rather than cutting down the trees, the partnership between the Ceiba Foundation and the Lima family aims to enable the family to generate an income from the living forest in such a way that ecotourists and scientists alike have the opportunity to enjoy and study this unique habitat.

The EcoMinga Foundation
'The cloud forests of the eastern Andes of Ecuador are the richest in the world for orchids. Warm wet winds blowing from the Amazon basin caress these mountains and are forced upward, cooling and releasing their moisture as they rise. The moisture condenses into nearly permanent fog blankets, covering the mountain peaks so thickly and so consistently that some of them have never been mapped, and appear on topographical maps only as mysterious white holes labelled 'no aerial photos available.' In this nurturing environment tiny, delicate orchid species have evolved, species with flowers so fragile they would collapse in minutes in an ordinary environment. Most of these are from the Pleurothallid subtribe: Lepanthes, Stelis, Platystele, and many more.' Lou Jost

The EcoMinga Foundation was established in 2005 by an international group of conservationists and scientists who were concerned about the rapid disappearance of Andean forests. It currently manages eight reserves in Ecuador and two reserves in Colombia within the Andean biodiversity hotspot.

Lou Jost, an ecologist and founder of EcoMinga, tells us that patterns in the diversity of orchids have become apparent when he walks in the forest in the region around his home in Baños, near the Tungurahua volcano in Ecuador. In the past three years, Jost and his students (especially Andy Shephard) have discovered about 28 new species of *Teagueia* in the surrounding mountains, with the greatest species density found on the east-facing slopes. Around 30 species of *Teagueia* occupy a small region of the Upper Río Pastaza watershed, some peaks harbouring 15 or 16 species. For instance, 16 species have been found on a single mountain – Cerro Candelaria. How did that happen? It would seem likely that this explosive speciation in the upper watershed of the Río Pastaza is tied to variations in the microclimate. Similarly, the mountains are home to a high diversity of *Lepanthes* species, some of which appear to exhibit extreme habitat specificity. Jost gives an example where a species of *Lepanthes* can be found along a mountain chain running north to south, but the same species seems unable to bridge the gap from west to east between parallel mountain chains, even though the distance is shorter.

On many reserves, a human presence is essential to deter poachers and prevent squatters setting up home within the conservation area. These 'keepers of the wild' are vital ambassadors for conservation within local communities and, given enough time, elicit the support of that community. This is not a problem in the area around Baños because it is a centre for ecotourism, and many people in the town now earn a living from the preservation of biodiversity. Where people are poorer, as in western Ecuador, there is continuing deforestation and mining for gold. Whenever gold is discovered, there is always a little gold rush, and environmental destruction soon follows.

The Dracula Reserve is a conservation area in north-western Ecuador, near the border with Colombia. This is a lush green landscape of forests and mountains currently covering over 1,133 hectares (2,800 acres) and is the centre of diversity for the genus *Dracula*, for which the reserve was named. Biological field research is being carried out here by several national and foreign institutions, and there are sure to be more exciting discoveries as they explore this incredible place. But today, the very survival of the reserve is threatened by illegal mining, serving as a reminder that reserves require resources for enforcement if they are to remain.

Indigenous reserves

As we have seen with *Laelia dawsonii* f. *chilapensis* in Mexico (see page 162), indigenous people can have an important role in protecting rare species, and can help to safeguard orchid populations that grow in protected natural reserves that have been set aside as communal areas.

In Costa Rica, *Cattleya dowiana* grows in the primary forest along the Caribbean side of the Cordillera de Talamanca, mainly in indigenous territories and in the La Amistad International Park. In the 10,000 hectares (25,000 acres) of pristine forest in the Hitoy-Cerere Biological Reserve, the species is still relatively frequent on the Caribbean slopes of the Talamanca range. The borders of the reserve are surrounded by immense indigenous reserves that extend to the summit of the Cordillera. Since the beginning of the 20th century or earlier, indigenous populations have been collecting the plants for horticultural purposes. The available data suggest that this species is not rare and that viable populations are under protection in areas specifically designed to conserve biodiversity. In the past, indigenous people exchanged sacks of the plants for commodities such as clothing or food, and this kind of trade may still be happening today among collectors and illegal nurseries. The wild areas where *C. dowiana* grows are not cut down because they are reserves, but illegal extraction continues to be a problem. Recently, many plants have been reproduced in laboratories and sold at exhibitions, which can reduce harvesting pressure. Growing *C. dowiana* is not easy, however, and many plants are killed by inexperienced growers.

A wider perspective

To conserve orchids, it is necessary to protect not just the forest, but all components of this complex and megadiverse ecosystem.

A constantly wet and humid Ecuadorian cloud forest is home to many tiny orchid species.
Photo: Lou Jost

Camera traps at the Colombian Orchid Society (COS) reserve in Antioquia have revealed the presence of more than 45 mammal species, including pacaranas (*Dinomys branickii*), which are beaver-sized rodents that are threatened with extinction, puma (*Felis concolor*), ocelot (*Leopardus pardalis*), and brocket deer (*Mazama* sp.). Both the COS reserve and the El Pahuma Orchid Reserve are home to Andean spectacled bears (*Tremarctos ornatus*). South America's only bear species, it inhabits mid- and high-elevation cloud forests of the Andes, where it feeds on a wide variety of plants and fruits, especially the tender hearts of bromeliads – a high-fibre diet not dissimilar to the one preferred by China's counterpart, the giant panda (*Ailuropoda melanoleuca*). The main threat to the bear is habitat loss, although in part of its range it is still killed for its gall bladder, which is used in traditional Chinese medicine.

Not only are Colombia and Ecuador rich in orchids, they are also 'honey pots' for ornithologists. Roughly 170 bird species have been recorded to date at the COS reserve, including the spectacular black-and-chestnut eagle (*Spizaetus isidori*). At El Pahuma, visitors may see the plate-billed mountain-toucan (*Andigena laminirostris*) with its indigo breast, red rump, and yellow patch on its bill, and the Andean cock-of-the-rock (*Rupicola peruviana*), the males of which are noted for their beautiful scarlet or orange plumage, and communal courtship displays. Like toucans, the cock-of-the-rock has a diet primarily consisting of fruit. In common with

(left)
Lepanthes wakemaniae can be found growing in the cloud forests of northern Colombia.
Photo: Sebastian Vieira

(opposite)
'Guaria de Turrialba', *Cattleya dowiana* is one of Costa Rica's most beautiful native orchids.
Photo: Diego Bogarín

most medium to large birds (the exception being parrots, which can digest seeds), seeds pass through its digestive tract largely intact, which makes the birds important dispersers of tree seeds throughout the forest, indirectly benefiting the epiphytic orchid flora.

EcoMinga's Dracula Reserve is an example of an area that contains an incredible diversity of flora and fauna, many of which are endemic to the valley in which the reserve is found. Early surveys identified over 160 different orchid species from 55 genera, and this list continues to grow. In addition, recent discoveries include two new mouse species (*Pattonimus ecominga* and *Chilomys georgeledecii*), shrew opossums, new species of frogs (*Hyloscirtus conscientia* and *Pristimantis* species), toads, and lizards (*Anolis dracula* and *Echinosaurus fischeri*). The reserve also contains new orchid species, including *Scaphosepalum zieglerae*, *Pleurothallis chicalensis*, *Trevora* species, *Pseudolepanthes bihuae*, and *Lepanthes tulcanensis*, as well as several new *Lepanthes* and *Pleurothallis* species that have yet to be described.

RESERVES

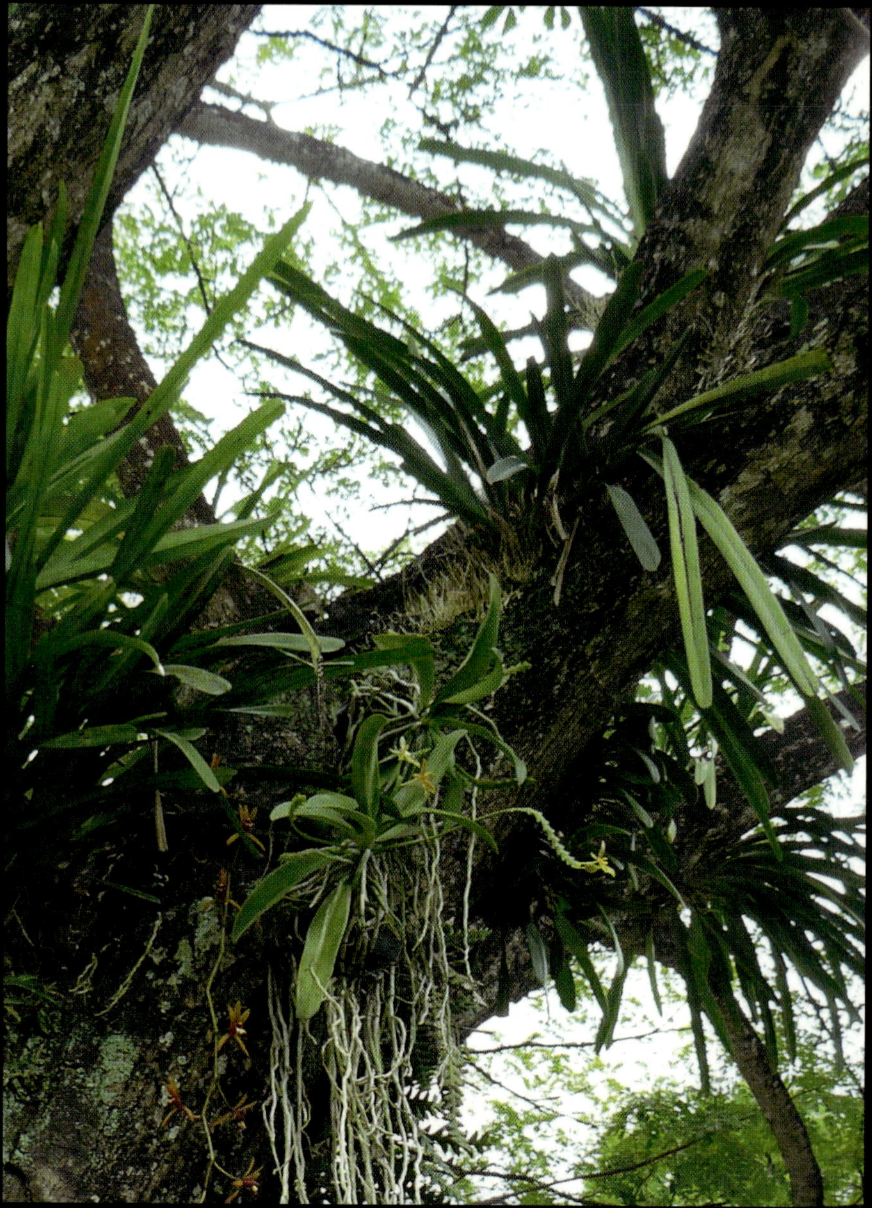

15 Reintroduction, restoration, and enhancement

Once primary forest is lost, it takes a considerable amount of time for healing and regeneration to occur through ecological succession: the process by which the structure of a biological community changes and evolves over time, until a mature forest (a 'climax community') is reached. Even if primary forest is eventually 'restored', it will differ from the original forest in the composition of its life forms.

Until now, the focus has been on maintaining individual populations on the basis that they may contain genetically distinct races or forms (ecotypes) adapted to their specific local environments. Perhaps the time has come to question this strategy. Why? Because, as climate changes, becoming warmer (or in some areas cooler), and wetter or drier depending on the locality, we may expect some isolated populations to begin to decline. In addition, habitat fragmentation hinders the ability of orchids to colonise new sites by reducing the number of seeds that are dispersed into suitable locations that harbour the mycorrhizal fungi needed for seed germination. As a result, artificial propagation and translocation of plants may become necessary. Introducing favourable genes found in plants from different populations (assuming the species found is accurately identified) may increase the ecological fitness of a declining population, thereby increasing its resilience in the face of climate change. Such an approach brings its own risks, principally the possibility of introducing pests or viruses. This can be overcome by raising plants from seed with strict quarantine conditions applied to ensure that they are not contaminated before reintroduction.

Reintroducing orchids in a private reserve in Brazil
In 2018, Luciano Zandoná and his wife, Angélica Guidoni Maragni, purchased 8,000 square metres (2 acres) of land next to the Cantareira State Park, with the aim of establishing a private reserve that would become a natural laboratory for studying orchids and other epiphytes. The neighbouring state park is home to 160 orchid species. By contrast, the private reserve is very poor in epiphytes because in the past the area was degraded by logging, and subsequently converted into coffee and tea plantations. Only eight epiphytic orchid species remain.

Zandoná selected 11 species that were native to Atlantic Forest in the metropolitan area of São Paulo. These were then grown from seed in a shade house. In September 2022 – springtime in the Southern Hemisphere, and the beginning of the rainy season in that part of Brazil – 1,200 seedlings of 11 species were tied to 100 trees, 12 seedlings per tree. Late secondary growth or climax trees were chosen – trees that were going to live a long time – with a corrugated bark that would allow orchid roots to better colonise (grip) the surface and not be shed.

The main objective of the project is to increase the population

(previous pages)
Orchids such as *Cymbidium finlaysonianum* and a (flowering) *Phalaenopsis cornu-cervi* (left) and *Bulbophyllum medusae* (right) have been introduced into a number of reserves in Singapore.
Photos: Tim Wing Yam

density of species that occur nearby, yet do not occur in the areas where the seedlings are being planted; thereby creating a gene bank of the orchids native to Cantareira. Plants are monitored monthly for signs of rooting, flowering, and production of fruits. The long-term success of any reintroduction programme ultimately depends on whether or not new seedlings become established, and the population then begins to increase. This in turn depends on the presence of suitable fungi either in the soil or, in the case of epiphytes, on the bark of the new host trees. Colonisation by fungi is currently being studied in collaboration with Samantha Koehler and her students from the University of Campinas. The aim is to establish the best protocols for the restoration of forests that contain orchids. Seeds of terrestrial species that occur in Cantareira, but not in the private reserve, are also being sown on the forest floor.

Cantareira State Park, where the work initially began, now has many new seedlings establishing themselves around the reintroduced plants and in the surrounding areas, where Zandoná began reintroductions 8–10 years ago, including that of the IUCN Red-listed *Oncidium hians*. It is still too early to detect seedlings in the private reserve, but they are expected to appear soon.

Saving Florida's cigar orchid
'This hammock is the most natural orchid garden in the United States. All but one or two of our native epiphytic orchids grow here . . . in greater profusion and to greater size than I have ever seen them elsewhere.' John Small talking about a coastal hammock in Everglades National Park in the 1920s

A 'hammock' is an island-like dome of trees on higher ground surrounded by a large expanse of a swampy area.

Around 1988, John Hall photographed one specimen of a cigar or cowhorn orchid (*Cyrtopodium punctatum*) on which he counted 31 four-feet-long flower spikes supporting hundreds of honey-fragrant blossoms. By 2007, fewer than 20 individuals were known to occur in Florida's Fakahatchee Strand. Loss of habitat due to intensive logging of its large stands of bald cypress for timber in the 1940s, combined with harvesting by over-zealous collectors, had decimated populations of the orchid. Although still encountered in the adjacent Big Cypress National Preserve, Florida Panther National Wildlife Refuge, and Everglades National Park, it only occurred sporadically. With no natural seed set or seedling recruitment documented in its recent history, the prospects for recovery of the species in the Fakahatchee Strand seemed dismal. Enter Atlanta Botanical Garden (ABG).

Members of ABG's recovery team began cross-pollinating individuals within the Fakahatchee Strand Preserve State Park, and in 2009 the first capsule was harvested and sent to the garden's

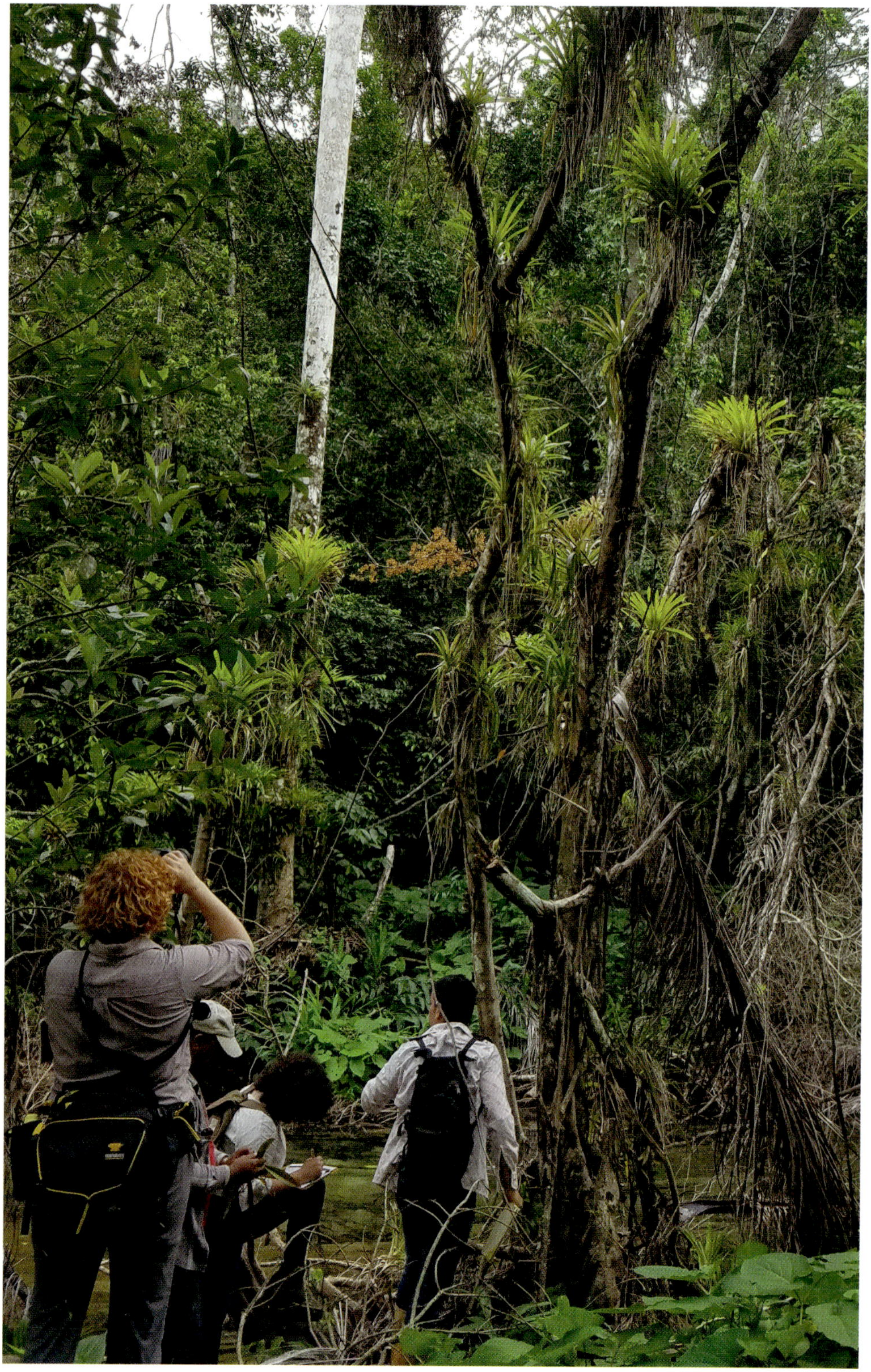

tissue culture lab, where staff and volunteers propagated plants for recovery. In spring 2011, a Brazilian student, Danielle Ferreira, visited ABG to participate in the conservation programme. Ferreira found that germination was marginally better in the dark than in the light, and that the magic ingredient for both germination and subsequent growth was the incorporation of banana into the culture medium. The addition of pulped banana is not suitable for all epiphytic orchids, but for *Cyrtopodium punctatum*, its benefits were clear. Eventually hundreds of seedlings were successfully transferred from the laboratory to the production greenhouse.

In the spring of 2011, the ABG team returned to the state park to spend four days in the field. The group carefully scouted for and selected microsites that would provide conditions as close as possible to those thought to be required by *Cyrtopodium punctatum*. Although the cigar orchid is dormant over the dry winter months – when it loses its leaves and relies on its long, plump pseudobulbs to survive desiccation – establishing the orchid was not always a given. During April/May, it begins to produce new roots, and develops showy flowers before the production of new growth. Seed capsules take 14 months to develop, and seeds are shed in the hot and humid months most conducive to germination. During their visits to the site, the team harvested a proportion of the capsules and took them back to the lab in Atlanta to ripen. Around 10% of the seeds were sown or put into dry storage at $-20°C$ ($-4°F$). The remainder of the seed was sent back to the Fakahatchee, where park rangers played 'Johnny Orchid Seed', scattering the seed at suitable sites.

The long-term aim of any re-introduction project is to establish a self-sustaining population that contains genetically distinct individuals. When the team discovered a clutch of new seedlings from seed dispersed in the previous spring of 2010, there was reason for hope. In 2011, 88 robust seedlings were introduced back into their natural environments throughout the reserve, more than quadrupling the known number of individuals throughout its native haunts within the park. Ferreira then cross-pollinated a selection of different individuals to increase genetic diversity, with future seed collections in mind. Fruits pollinated in the previous year were collected for *ex situ* production at ABG, and further seed was sown *in situ* at appropriate microsites.

During the past 10 years, scales and other insect pests have infested *Cyrtopodium punctatum* and other native epiphytes throughout the Fakahatchee Strand. Many of these pests are thought to be invasive. Much of south Florida has been riddled with many different kinds of invasive species, such as monster reticulated python (*Malayopython reticulatus*), Brazilian peppertree (*Schinus terebinthifolia*), and the Mexican bromeliad weevil (*Metamasius callizona*), which has decimated the bromeliads in the region. How did these pests get here? Nobody knows for sure, but

An expedition by members of Atlanta Botanical Garden and Cuban colleagues looking for the cigar orchid (*Cyrtopodium punctatum*).
Photo: Atlanta Botanical Garden

(oppositte)
Danielle Ferreira hand-pollinating *Cyrtopodium punctatum* in the Fakahatchee Strand, Florida.
Photo: Matt Richards

(above)
A magnificent cigar orchid (*Cyrtopodium punctatum*) photographed by Roger Hammer in 2016. It is clinging to its long-dead host tree – a reminder of old Florida.

(left)
Sadly, the orchid is no more, a victim of Hurricane Irma in 2017.
Photo: Roger Hammer

the most likely source is through ports in Miami to the east. With each passing year, the invasives become more numerous, and new ones are discovered, adding an already expensive burden to the land managers in the region. It is too early to know how 'natural' south Florida's unique ecosystems will remain, and how much they will still resemble those described by Marjory Stoneman Douglas in her book, *The Everglades: River of Grass* (1947). But one thing is for sure, south Florida needs help. On the bright side, many of the same species found in south Florida are also present in Cuba, including *C. punctatum* in the Ciénaga de Zapata Parque Nacional. Despite a lingering political rift between the USA and Cuba, researchers and garden staff at ABG and in Illinois are able to share their expertise and work alongside Cuban scientists to save the species they share.

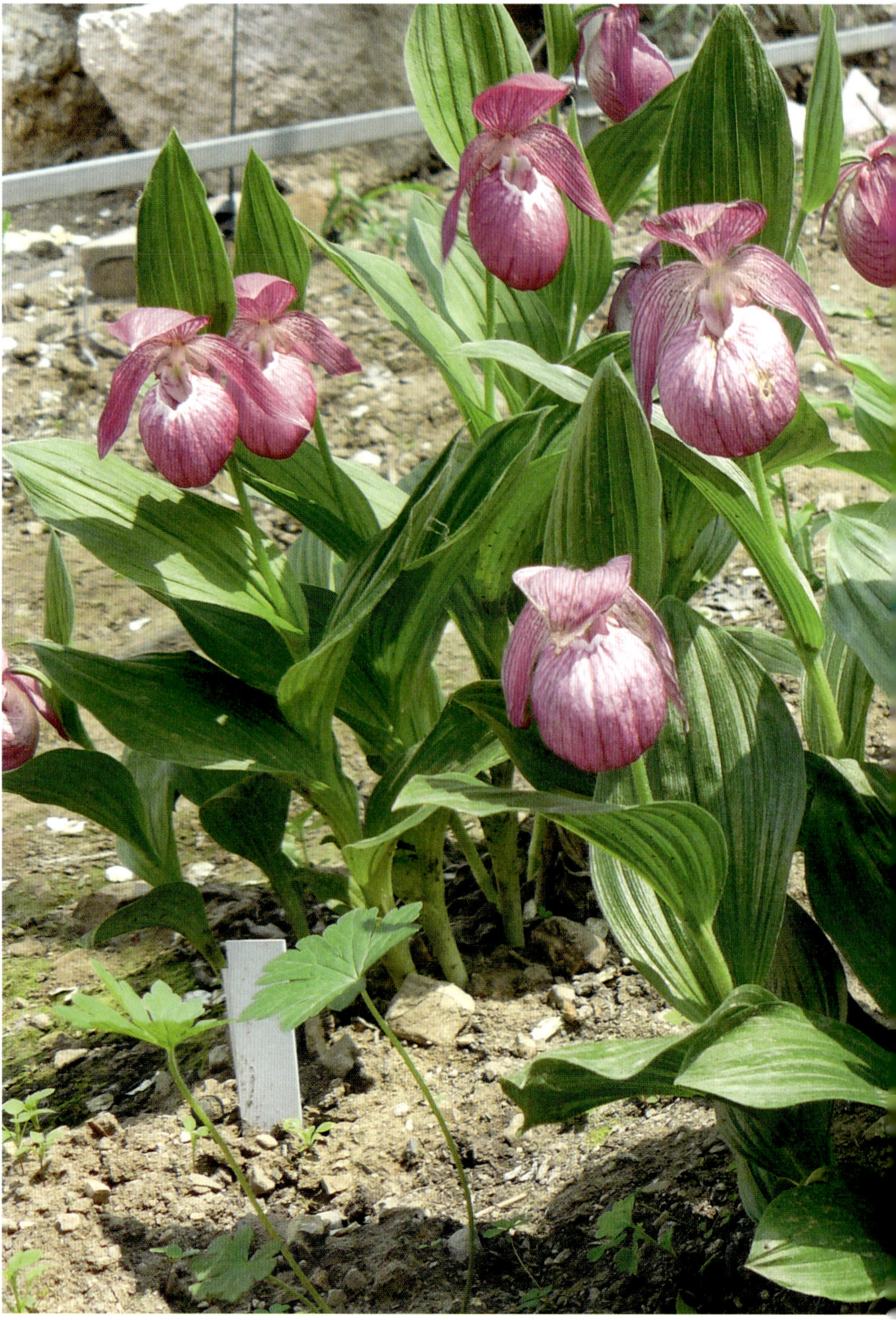

16 Slippers

Slipper orchids are different. The lip is modified to form a pouch: they are typically 'trap flowers'. As such, they have something in common with the North American trumpet pitchers (*Sarracenia* species) and the Asian *Nepenthes*, although they don't digest their prey. The word 'slipper' seems to be derived from the Old English word *slypescoh*, meaning 'slip-shoe', and refers to casual footwear worn indoors. It is applied to mostly terrestrial orchids in the genera *Cypripedium*, *Paphiopedilum*, and *Phragmipedium*, which are known for their large, often solitary flowers that resemble dainty footwear. The pollen consists of two glutinous masses located on either side of the column, and students often need convincing that the slippery disc hidden under the column and inside the pouch-like lip is where they should smear the pollen. To see the stigma, either the pouch must be removed, or a window cut in the back of the pouch.

With buds like birds' eggs that gradually unfold to reveal flowers of an ephemeral and fragile beauty, cypripediums are among the most desirable of temperate orchid species. All are endangered to a greater or lesser degree. Found in North and Central America, Europe, and Asia, recent assessments of their status by the IUCN Orchid Specialist Group, under the chairmanship of Kew's Mike Fay, make shocking reading. About 8% are classified as 'critically endangered', 46% 'endangered', and 25% 'vulnerable'. Another 9% are considered 'near threatened'. *Cypripedium dickinsonianum*, for example, is known only from a few scattered populations in Mexico, Guatemala, and Honduras. Its open forest habitat is being cleared for agriculture, and the cutting down of trees is changing the environmental conditions that allow these orchids and other understorey plants to thrive.

Native to tropical Asia, roughly 99% of *Paphiopedilum* species are likewise threatened with extinction, primarily due to over-collection for horticultural purposes and habitat loss. However, highly damaging illegal trade continues due to a lack of adequate enforcement of CITES regulations. Although these species are mostly well-represented in cultivated collections, their declining numbers in the wild are a cause for concern, because the consequent reduction of their genetic diversity threatens the orchids' continued existence in their natural habitats.

The story of *Paphiopedilum helenae* is all too familiar. 'For many years only a few species of the genus *Paphiopedilum* were known from Vietnam,' wrote the Russian botanist Leonid Averyanov in *Slipper Orchids of Vietnam*. In 1995, he discovered *P. helenae* near a mountaintop in a karstic limestone region close to the border with China. 'The fate of the species at this location was sad. Only two months after the publication of the description, orchid dealers engaged the locals to collect the plants. As a result, all plants, including seedlings, were completely collected in a few weeks.'

Before the discovery of the scarlet *Phragmipedium besseae*

(previous pages)
Cypripedium macranthos under cultivation in the mountains outside Beijing.
Photo: Philip Seaton

along a Peruvian roadside in 1981, the flowers of phragmipediums in cultivation were a subtle combination of browns and greens that lacked the visual appeal of the other slipper orchids. But the discovery of *P. besseae*, followed by another sensational Peruvian beauty, *Phragmipedium kovachii*, 20 years later, was a game changer. Not only were the scarlet blooms of *P. besseae* and the outrageously large purple blooms of *P. kovachii* highly desirable, but they also brought with them the possibility of breeding a spectacular range of vibrant colours of South American slippers into the commercial market.

Many orchids, but especially slipper orchids, seem to cast a spell over the beholder. Some people are so entranced by their beauty, they allow themselves to become enmeshed in a web of illegal activities, whether fuelled by commercial greed, desire, fame, or all three. Like the Florida ghost orchid, the saga of *Phragmipedium kovachii* became the subject of a book – *The Scent of Scandal*. With enormous purple flowers, *P. kovachii* was discovered in the mountains of northern Peru in 2001. The Peruvian authorities gave permission to three nurseries to each collect five plants in 2003. But within a few years of its discovery, thousands of plants had been illegally removed for sale on the internet for enormous sums – up to $10,000 for two plants. Would it be ethical for seed from a potentially valuable orchid to be sent to another country for commercial growers in foreign countries to profit from? Of course not. Thus, clearly written rules and regulations are needed for both the amateur and commercial grower. But enforcement of these rules and regulations will only go so far, as there is little to stop nurseries in another country from growing seedlings for profit or generating new hybrids from selective breeding programmes.

A fourth slipper

The story of *Mexipedium xerophyticum* demonstrates the extreme vulnerability of some species in their natural habitats, and the pivotal role that *ex situ* techniques can play in conservation. Discovered in 1988, *M. xerophyticum*, which is the sole member of the genus, is endemic to a warm and humid region in Los Chimalapas, in the state of Oaxaca, Mexico. At the time of its discovery, just seven 'genets' (colonies of genetically uniform and distinct plants) were found growing in pockets of vegetation on rocky limestone outcrops, immersed in a matrix of tall evergreen forest. The location was difficult to access, and a large part of the surrounding area had been transformed by human activity. The plants spread vegetatively by means of stolons, and there was no evidence of natural recruitment of seedlings. Thus, it seemed unlikely that such a small population would persist in the wild. One complete genet and part of another were removed and given to botanical gardens and specialist growers for propagation. These plants subsequently became the source of material that is currently

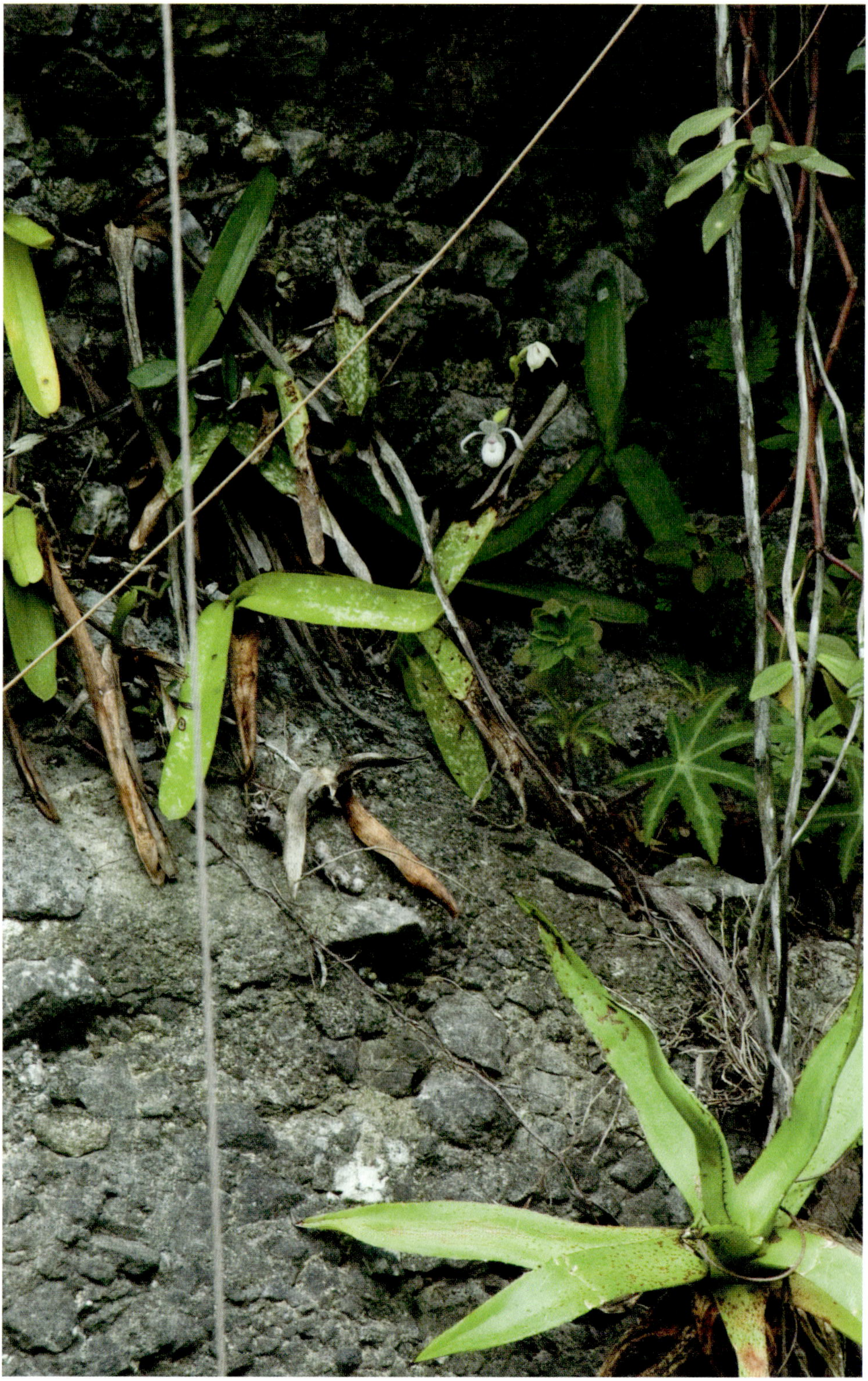

in cultivation worldwide, both by vegetative propagation and from seed. A subsequent visit by the scientists to the (secret) site, revealed that two more genets had been removed by a person or persons unknown.

In July 1996, two more genets were found. The precarious situation of the six remaining genets became critical when a fire destroyed the habitat in 1998, and an expedition in 2009 found just one surviving plant, severely damaged, but recuperating. Further exploration over difficult terrain revealed a few more plants growing on the vertical walls of a ravine. Exactly how many genets there are, however, awaits further molecular analysis, and a long-term study of the population dynamics is needed. Although their long-term future in cultivation is probably assured, the remaining plants in the wild are found in an area of less than two hectares (4.9 acres), and their future remains uncertain.

A golden slipper
The large golden flowers of *Paphiopedilum armeniacum* created a sensation when the orchid world became aware of its existence in western Yunnan, China in 1982 (it also occurs in neighbouring Myanmar). Its fate seemed to be sealed, however, as it was soon subject to over-collection and appeared for sale in large numbers, rapidly becoming a must-have species for impatient collectors. The story was a familiar one: poor people collecting the orchids for small sums of money to supplement their meagre incomes, and middlemen making large profits. Over-collection, combined with habitat loss and degradation, meant that the future prospects for the remaining and fragmented wild populations seemed dim indeed. What could be done?

Villagers were enlisted to help reduce collection pressure on *Paphiopedilum armeniacum*, and they were provided with alternative sources of income while doing so. To ensure maximum genetic diversity, researchers then collected plants from eight separate locations and took them to the National Orchid Conservation Centre in Shenzhen, where they were grown and pollinated by hand. The seed was later sown and, once they had reached adult size, the seedlings were successfully reintroduced into the eight natural sites. After a few years, several of these plants flowered, and new plants soon began to appear. Growing and successfully reintroducing slippers is no easy task, and although *P. armeniacum* is a small case study, the message it conveys is a powerful one. Using an integrated approach that incorporates both *ex situ* and *in situ* techniques, it is possible for scientists and local communities to work together for the greater good.

Mexipedium xerophyticum growing in its natural habitat.
Photo: Eduardo Pérez

English lady's-slippers
The golden yellow pouch and twisted claret petals of the English lady's-slipper orchid, *Cypripedium calceolus*, appear outrageously

VIEWPOINT

Orchid conservation at Kew
by Phillip Cribb

Shortly after I joined the staff of the Royal Botanic Gardens, Kew, I was asked to serve on a committee of 12 people whose purpose was to protect the lady's-slipper orchid, Britain's showiest native orchid. Gren Lucas, deputy director of Kew at the time, called it the 'giant panda of the plant kingdom'. As such it was a flagship species, whose successful protection could be used to protect less showy orchids and other plants. Having been considered extinct for several years, it was rediscovered in the 1930s, and eventually became the first plant in Britain to have its own Species Recovery Programme – a future model for other endangered plant species.

Thus, I became part of a long tradition at Kew of staff involved in practical plant conservation. The Lady's Slipper Orchid Committee, comprising representatives of the County Wildlife Trusts, local universities, Natural England (as it now is), and Kew, was set up by Edgar Milne-Redhead, former deputy keeper of Kew's herbarium. At my first meeting in the early 1970s, I was shocked that the main strategy for protecting the last remaining wild plant was to cut off all its flowers so that it would not be noticed by walkers in the Dales. Fortunately, subsequent discussions led to a more proactive approach of attempting to propagate it from seed, with the aim of reintroducing it in the future.

One of the surprises I had when I started at Kew was just how long-lived orchids can be. Several orchids growing in the T-range greenhouses at Kew had been donated in the 19th century; notably, gifts from the famous orchid nurseries of Fred Sander, Stuart Low, and James Veitch. This longevity boded well for our native lady's-slipper orchid.

As a taxonomist specialising in orchids, field work around the world has provided me with a broad appreciation of the biology, ecology, and conservation of orchids. Accurate identification and naming of orchids is basic to orchid conservation. Good field guides or floristic accounts exist for orchids in many countries around the world, but there is still a lot that is poorly known or even unknown. In many species-rich countries in the tropics, such as New Guinea, Sumatra, Colombia, Ecuador, and Peru, identification remains tricky – at times exceedingly so.

Even in the 1970s, the orchid growers at Kew were harvesting orchid seed from plants in the collection and storing it at 4°C (39°F) in a refrigerator, often for many years. At about this time, Peter Thompson, Kew's seed physiologist, started a micropropagation programme, and successfully grew some of this seed. He later moved the programme to Wakehurst Place, where he was succeeded by Hugh Pritchard in 1982. Vials of seed from all of Britain's orchid species and hundreds of tropical orchid species are now successfully stored there – an essential reserve against extinction. Over the past few years, Phil Seaton has successfully taught techniques and established orchid seed storage and propagation in a number of countries that previously lacked the expertise. Nowadays, the collection and storage of orchid seed is a standard procedure worldwide, useful to conservation, research, and horticulture. For the past 30 years, Lawrence Zettler has been isolating and using mycorrhizal fungi to germinate orchid seeds from North America, Hawaii, the Republic of Palau in the Pacific, and Madagascar. His work in Madagascar was in collaboration with my colleagues at Kew, and also at Kew's Madagascar Conservation Centre – a facility that I helped to develop.

Although over-collecting has been a major threat to orchids in the wild, even greater threats are posed by habitat neglect and destruction. These are now compounded by climate change. Many orchids cannot adapt to a rapidly changing climate and the consequent dangers of excessive temperatures, drought, flood, and fire. Managing habitats to preserve biodiversity is practised on a wide scale by government nature protection organisations, conservation NGOs, and by small groups and individuals. I have had the privilege of being involved in several hands-on projects to improve orchid habitats in Britain and in continental Europe. Sammy Sprunger, an ex-Kew student from the 1960s, has organised several habitat improvement projects in the Jura in Switzerland and Alsace in France (see page 240). In the Jura, he organised the clearance of a 20-hectare (49-acre) plot overgrown with the invasive tree *Robinia pseudoacacia*, where he had found a few non-flowering rosettes of *Ophrys* on a walk. Thirty years later, the site boasts over 25 species of orchid, some present in their thousands. Nearby, a two-kilometre strip along the fence of Basel airport was cleared, and now has a rich fauna and flora that has proved to be a wonderful study site for undergraduates from a nearby university. More recently, he organised a three-year project to reintroduce the lady's-slipper orchid to many sites in the Swiss Jura. The lesson is clear: individuals can make significant contributions to the conservation of plants, including orchids.

In the late 1980s, the establishment of the IUCN Species Survival Commission's Orchid Specialist Group linked orchid researchers and conservationists around the world. The exchange of expertise and personnel has proved to be a significant spur to developments in orchid conservation. In 2003, as chair of the Orchid Specialist Group, I teamed up with Kingsley Dixon and Russell Barrett in Perth, Western Australia, and Shelagh Kell at Kew to set up the first International Orchid Conservation Conference in Perth. We also co-edited *Orchid Conservation* (2003), a handbook of current knowledge, methodologies and techniques. The three-yearly conferences continue to provide invaluable opportunities for networking. For example, several groups heard about Hanne Rasmussen's pioneering work on recovering mycorrhizal fungi by using 'seed slides' planted out in nature (see page 38). Using the right mycorrhizal partner can be critical to the success of orchid propagation and reintroduction efforts. Another benefit of the international profile that orchid conservation has achieved is the influx of young and enthusiastic talent to the conservation movement.

---

exotic in a northern English woodland; but its beauty almost led to the species' demise. In the 19th century, lady's-slippers were collected as desirable garden plants and as herbarium specimens. By 1917, the species had been declared extinct in England. However, in 1930, a single plant was discovered in Yorkshire. Its location remained a closely guarded secret for many years to preserve it from unscrupulous collectors.

But in 1966 disaster struck. The orchid's guardians reported a hole in the ground where the plant had once been rooted. Fortunately, not all of the rhizome had been dug up by the culprit or culprits, and the following year four shoots appeared. Action was clearly needed to prevent the loss of this lonely iconic plant, and in 1970 the *Cypripedium* Committee was formed with the aim of ensuring its safety. Henceforth, the plant was kept under constant surveillance by a guardian during its growing and flowering season. The number of flowering shoots increased over time and artificial pollination from the early 1970s onwards led to the production of seed for *ex situ* cultivation at both the University of Leeds and Kew.

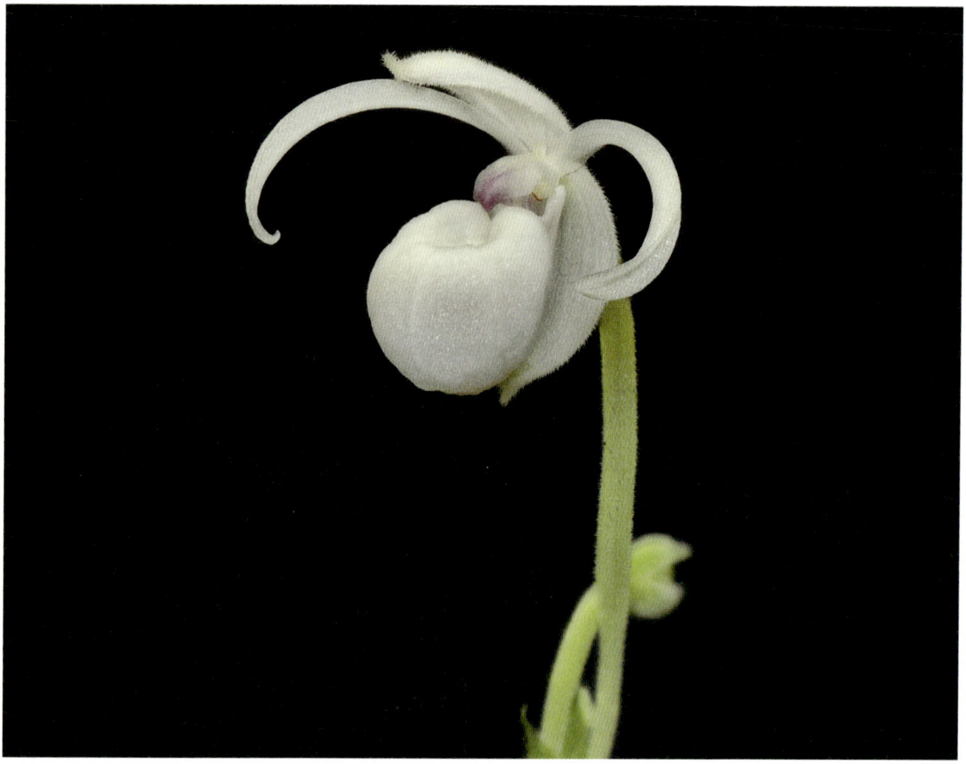

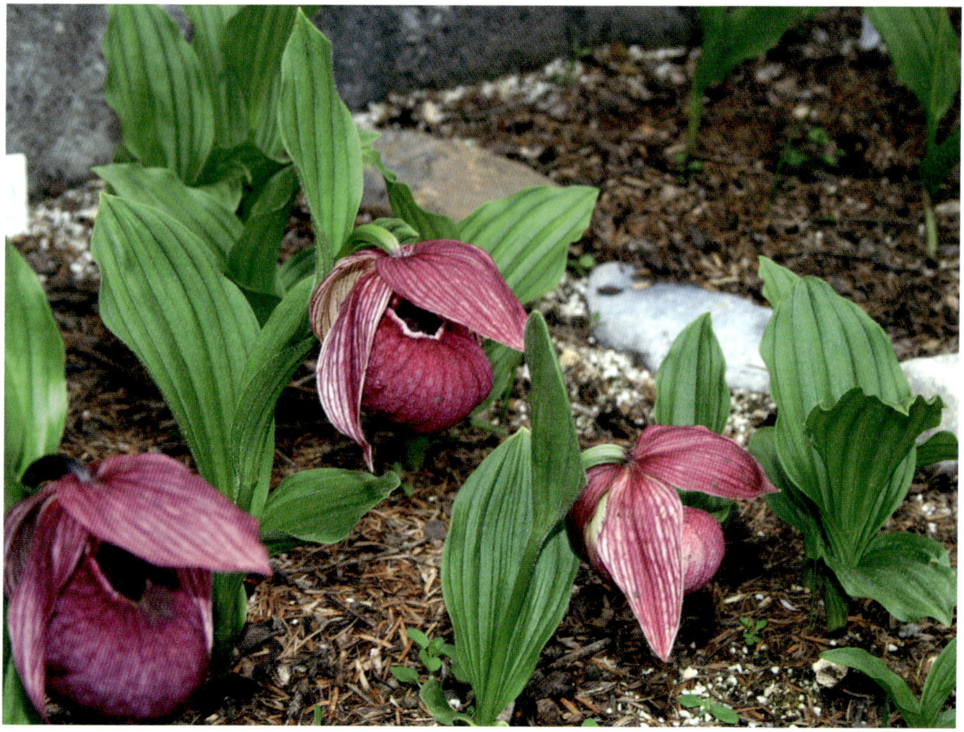

(opposite top)
Mexipedium xerophyticum close-up.
Photo: Mike Bull

(opposite bottom)
Cultivated plants of *Cypripedium tibeticum* in Huanglong.
Photo: Holger Perner

A few seedlings resulted from scattering seeds around the surviving plant, two of which eventually flowered. Natural pollination eventually occurred in 1999, yielding three seed capsules. A portion of the seed was sown around the base of the plant in the hope that some would germinate naturally. The rest was taken back to the lab to attempt *in vitro* germination.

'Porridge-eating fungus saves endangered orchids', declared an article in *New Scientist* in May 1983. Phillip Cribb had obtained funding, from Sir Robert and Lady Sainsbury, for a project to develop protocols for propagating British and European orchids whose numbers in the wild had been greatly reduced so that they could be reintroduced. Cribb invited Mark Clements over from Australia to work in Kew's Micropropagation Unit. Clements was known for successfully growing endangered Australian native species for reintroduction from seed, using mycorrhizal fungi. Fungal isolates were grown on a simple 'oats medium' – a recipe based on powdered porridge oats, with agar added as a setting agent. Success with a range of European orchid species soon ensued and Clements successfully grew several rare species. Three species, including the lax-flowered orchid, *Anacamptis laxiflora*, were introduced to Wakehurst Place in the Sussex countryside (now home to the Millennium Seed Bank) in 1984, where they have survived ever since.

Unfortunately, the lady's-slipper resisted all attempts to yield a fungus from its roots that would lead to symbiotically grown seedlings. Indeed, isolating suitable fungi from any *Cypripedium* species has so far eluded mycorrhizal scientists, and not through lack of trying. Clearly a 'porridge-eating fungus' was not the answer. The embryos of *Cypripedium* seeds are enclosed in a hard, black carapace and, in nature, go through a period of winter dormancy during which the carapace is broken down. Although it is possible to break down the carapace using chemicals, the best method turned out to be to sow immature seed before the carapace had developed fully, and to use Malmgren's asymbiotic medium (see page 283). Using this technique, Robert Mitchell and Margaret Ramsay began to grow seedlings from immature capsules. But Kew didn't have the facilities to grow large numbers of young plants in compost, so Cribb approached Kath Dryden of the Alpine Garden Society, who assembled a group of their members who had successfully grown *Cypripedium*. They were asked to take Kew's surplus seedlings and grow them as a reserve stock in case the reintroductions failed.

After the plants had been grown to a suitable size, they were planted in experimental plots to monitor their survival. Perhaps unsurprisingly, it was found that large plants fared best. By 2003, 1,500 seedlings had been planted out in 14 different locations, including one accessible to the general public. Despite the many dangers and predations that often threaten young plants, there are now several thriving lady's-slippers at a number of sites and several have since flowered. Then in 2009, seed capsules formed after natural pollination. The real measure of success will be the appearance of lady's-slipper orchid seedlings in a new location. Such an event would suggest that a fungus required for seed germination remains in the soil at the site, which could be isolated and added to seeds, thereby generating symbiotically raised seedlings for reintroduction.

How will populations be impacted by future climate change? In 2022, it became apparent that the long-term survival of many of the lady's-slippers that had been planted in public locations was in doubt due to the increased frequency of droughts. Without watering and protection from slugs, it was considered unlikely that the plants would thrive and recolonise naturally. Plants in a shadier, damper and more vegetated area flowered well in both 2019 and 2020 without intervention, indicating that this type of location may be more suitable for future introduced plants. However, as weather patterns change, there will be a need to identify places where plants will thrive in the future with minimal intervention.

A specific aim of the project was to introduce specimens to sites where they could be enjoyed by the public. A vigorous specimen of the lady's-slipper at Silverdale in the Gait Barrows National Nature Reserve in Lancashire, which is thought to have been planted in Victorian times, has been cared for over many years by dedicated local naturalists. Additional plants were introduced with the goal of making the reserve a 'honeypot', allowing controlled public access to plants without adding pressure to the wild sites where re-introduced plants were in their most fragile stage of growth. However, using DNA fingerprinting techniques, it was found that all of the plants at Gait Barrows had been derived from a plant that originated in continental Europe. Did this matter? The *Cypripedium* Committee believed so. Having found that there was sufficient genetic diversity in purely English stock, it was the committee's opinion that re-introducing foreign genes carried with it an unnecessary risk of outbreeding depression (when crosses between two genetically distant populations result in a reduction in fitness). Consequently, all plants containing 'foreign genes' were removed from the programme and the re-introduction sites, including from Gait Barrows. They were dug up, and to this day are being cared for by committee volunteers.

Why bother to maintain these plants? After all, germinating and growing seedlings is expensive. *Cypripedium calceolus* is still widely distributed in Europe, Scandinavia, temperate Asia, and possibly Japan. However, although some large, healthy populations remain, the species is critically endangered in most of its European habitats. For example, when Phil Seaton visited the Bükk National

Peter Corkhill pollinating *Cypripedium calceolus*.
Photo: Phillip Cribb

Park in northern Hungary in 2015 to see C. *calceolus* growing in the woodlands, he was delighted to find that, as well as flowering plants, seed capsules and seedlings were present – only to be told by the ranger, József Sulyok, that the climate was changing in such a way that there were now droughts in the springtime. Seedlings soon succumbed and were not seen again. Not only that, but the larger colonies of C. *calceolus* were gradually diminishing in size, and the orchid would probably disappear from the region in the none-too-distant future.

Cypripedium calceolus in Switzerland

Although some healthy populations of *Cypripedium calceolus* still exist in the Swiss Alps, the same is not true in the Jura mountains, where most populations are severely depleted or extinct. Despite strict legislation, the illegal poaching of plants continues.

In 2013, a project to reintroduce *Cypripedium calceolus* in Switzerland was undertaken by the Swiss Orchid Foundation, orchid breeder Anthura B. V. in Bleiswijk, Holland, and the conservation authorities of nine Jura cantons. Plants were hand-pollinated, and detailed records kept. Pollination by hand has the advantage of allowing researchers to choose which plants to use as pollen donors and which to use as the 'mother plants'. The ability to transfer a larger pollen load to the stigma of the flower can also result in more seed, though the number of possible genetic combinations within a small population is very limited. Such a limited gene pool can cause 'inbreeding depression', resulting in poorly growing offspring and tremendous losses during artificial propagation and re-introduction into the wild. Without molecular data to assess the gene pool of a population, these negative effects only become evident when the seedlings of these plants are grown in the greenhouse. If inbreeding depression occurs, and is severe, that particular population will gradually dwindle over time, eventually becoming extinct as a result of the depleted gene pool. For this reason, it is important to understand the genetics of a rare orchid population poised for recovery before valuable resources are invested. In cases where an orchid population suffers from severe inbreeding depression, it is necessary to decide whether to introduce plants from existing populations, or whether to tap into the gene pool of nearby populations to improve genetic vigour.

In June 2018, 3,000 adult plants were relocated to 48 sites. A year later, 75% of the reintroduced plants had survived and developed well. The project was deemed a huge success. It serves as an example of a successful collaboration between government authorities, a non-profit organisation, and a private company to re-introduce and protect an endangered species in the wild. The collaboration demonstrated that a rare orchid species may benefit from reintroduction efforts while, at the same time, making sustainably produced high-quality plants available to amateur growers, which reduces pressure on wild populations.

Cypripedium macranthos in China

Once upon a time, large populations of *Cypripedium* species could be found growing in the mountains to the north-east of Beijing. Today, they have become increasingly rare and difficult to find. A study conducted by staff at Beijing Botanical Garden concluded that, unless action was taken, the prognosis for their long-term survival in the wild was not good. Human disturbance, including habitat destruction, tourism, over-exploitation, over-collection,

Reintroducing *Cypripedium calceolus*.
Photo: Samuel Sprunger

and grazing, were the major causes of the rapid decline. Despite their commercial sale being prohibited in China, a large proportion of Chinese cypripediums remain targets of plant collectors, and they have been smuggled out of the country to destinations in Japan, Europe, and North America. In 2008, the Beijing Municipal Government listed all three native species of *Cypripedium* (*C. macranthos*, *C. calceolus* and *C. shanxiense*) as endangered, requiring the highest level of protection.

    The orchid team at Beijing Botanical Garden is conducting in-depth studies into wild germplasm resources of cypripediums in north and northeastern China, as well as surveying and recording the population distribution and survival status of the species and their natural hybrids. Based on their research, they have developed techniques for artificial propagation of *C. macranthos*, combining *in situ* and *ex situ* conservation, to assist the recovery of wild populations. A nursery has been established in the mountains,

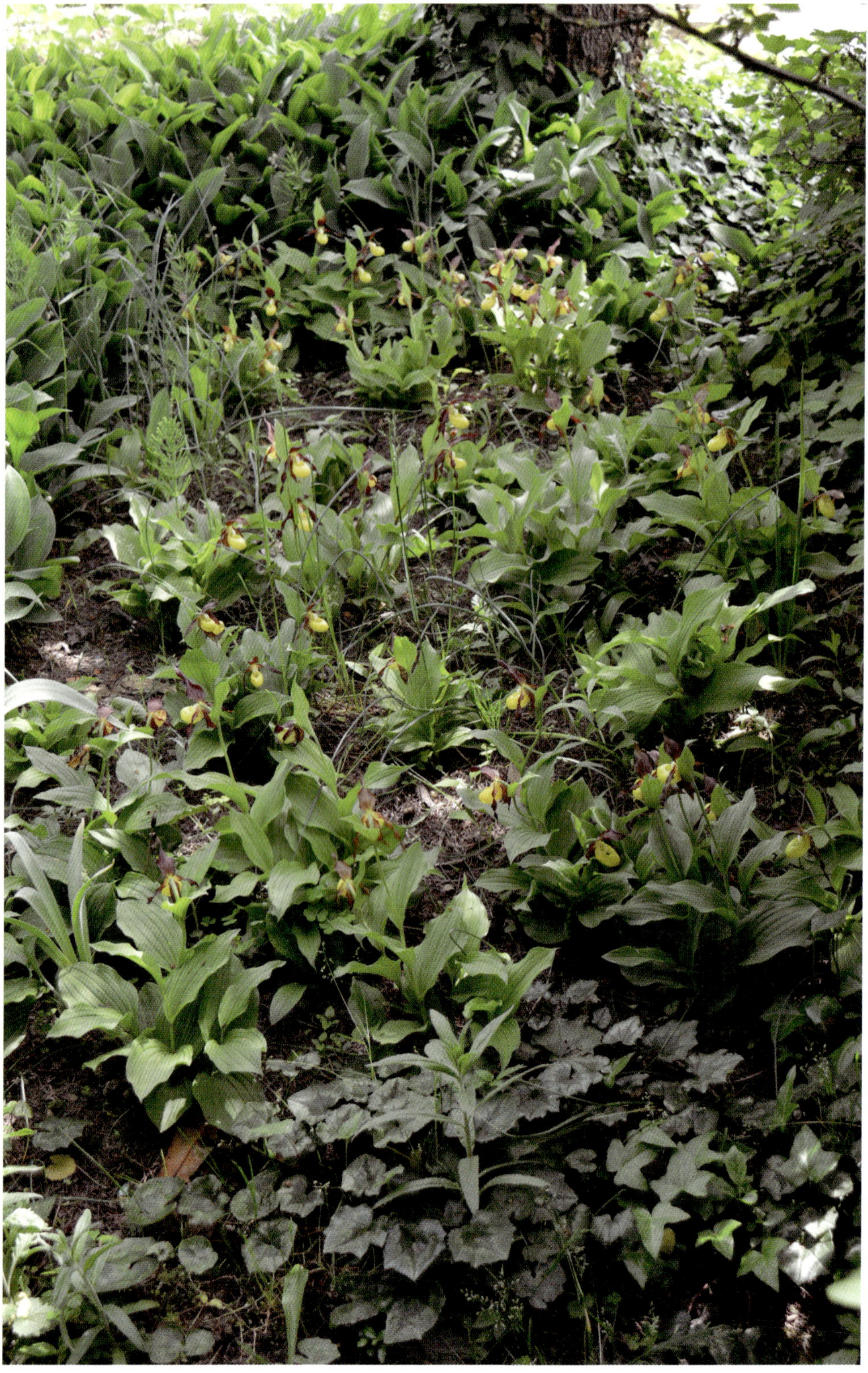

Reintroduced *Cypripedium calceolus* in bloom.
Photo: Samuel Sprunger

where the cool climate is amenable to *Cypripedium* culture. The minimum temperature is around –20°C (–4°F) in the winter, and generally below 30°C (86°F) during summer. The loamy soil in what was once a vegetable garden has since been enriched with some additional humus, and the pH adjusted to a little below 7 (slightly acidic). In addition to *C. macranthos*, the nursery contains small numbers of *C. calceolus* and *C. shanxiense*, plus *C. tibeticum* and *C. flavum*. Shelter from the summer sun is provided by shade cloth, which blocks 70% of direct sunlight. During the six months of winter, when the climate is dry, the shade cloth is removed and the plants are exposed to the elements. Apart from the occasional covering of snow, there is very little precipitation, and plants are given no protection apart from a little extra covering of soil for the newly transplanted seedlings. Plants from the nursery are reintroduced into their previous natural habitats at the beginning of the short rainy season at the beginning of June.

In 2016, Beijing Botanical Garden began cooperating with the Huanglong Nature Reserve, where there are 11 *Cypripedium* species. They have now successfully propagated several thousands of *Cypripedium* seedlings, such as *C. flavum*, *C. henryi*, *C. tibeticum* and *C. sichuanense*. In June 2019 and 2020, more than 2,000 young seedlings were returned to Huanglong. After transitional cultivation and acclimatising in the alpine conservation nursery for a couple of years, they will be introduced to the wild. Another success story involving a notoriously stubborn genus.

VIEWPOINT

My orchid journey – the origins of the North American Orchid Conservation Center
by Dennis Whigham

How does one become enthralled with orchids? There are many pathways. Mine began when I arrived at the Smithsonian in 1977 and started to monitor woodland herbs in the forests at the Smithsonian Environmental Research Center (SERC) near Edgewater, Maryland. Two of the first orchids I encountered were the cranefly orchid, *Tipularia discolor*, and putty root, *Aplectrum hyemale*. They captured my interest because they had produced leaves in the dead of winter and, as I later learned, there were no leaves in the summer. What was the deal? In the spring I enjoyed the beautiful flowers of a showy orchis, *Galearis spectabilis*, and noticed that the downy rattlesnake plantain, *Goodyera pubescens*, was evergreen. To book-end the diversity of orchid life histories in our forests, the autumn coral root, *Corallorhiza odontorhiza* – which is fully 'mycoheterotrophic', that is to say that it is entirely reliant on consuming fungi because it lacks chlorophyll – burst forth the first autumn I was at SERC.

As an ecologist interested in how plants 'make a living', orchids became a long-term interest. A few years later, we were fortunate that Hanne Rasmussen from Denmark came to the SERC lab and introduced us to the world of orchid mycorrhizal fungi. Rasmussen launched us in directions that have only become more intriguing now that molecular tools are available to ask additional questions. By that time, Melissa McCormick had joined

SERC, and I was increasingly in contact with others working on orchids around the world – including a sabbatical at Utrecht University where Jo Willems and I spent countless hours talking 'orchid' and visiting his study sites around the Netherlands. I then started to be more curious about orchid conservation work. Later, a visit to Perth, Australia, and Kingsley Dixon further whetted my appetite for orchid diversity and conservation. In short, I learned that while there are many wonderful efforts directed toward orchid conservation, few have had an ecological focus, incorporating the mycorrhizal fungi that orchids need to survive.

I thought an ecological approach would be valuable, but I wondered how to apply that perspective to orchid conservation. Fortunately, Bill and Melinda Gates gave the Smithsonian a grant that encouraged collaborative efforts through a competitive granting process. Bingo! A year later, I applied for seed funds and received additional funds to initiate an effort to conserve all of the native orchids in the US and Canada. This project was the birth of the North American Conservation Center (NAOCC), which has grown as a successful programme of SERC over the past 12 years, coordinated by Julianne McGuinness.

Why only the US and Canada? It was clear to me that success depended on working with a relatively small number of species. The US and Canada combined have just over 200 species, which was a number with which we could work. NAOCC integrates three interlocking components: first, to secure the genetic diversity of our native orchids and their mycorrhizal fungi; second, to establish a network of collaborations to support the collection efforts and to develop protocols for propagating native orchids; and third, to establish educational and outreach efforts to gain public support for orchid conservation and to enrich the lives of everyone with whom we engaged.

As I approach retirement, a few important objectives remain, but the NAOCC model – based on ecological principles – can be used anywhere in the world. To the many students, individuals, and organisations who have been our collaborators, my heartfelt thanks for being along with NAOCC for the ride!

English meadows and orchards
The UK has an orchid flora of around 50 species that can often be seen in large numbers in English meadows. Although a staggering 97% of those flower-rich meadows have been lost since the Second World War due to changes in agricultural practice, fragments of their past glories remain, mainly due to the efforts of county Wildlife Trusts and other voluntary bodies that have either purchased or leased the land from local farmers.

Eades Meadow is a reserve of approximately 6.5 hectares (16 acres) deep in the rolling Worcestershire countryside. Designated a Site of Special Scientific Interest (SSSI), it is one of the finest examples of lowland neutral grassland in England. It seems likely that much of the meadow has never been ploughed. Thus far, 130 plant species have been recorded on the site. Peering over the five-barred gate, as the seasons progress, visitors are rewarded with an ever-changing colour palette. In early spring, the meadow is a sea of the nodding heads of yellow cowslips (*Primula veris*), before it changes to shades of purple as tens of thousands of green-winged orchids (*Anacamptis morio*) begin to bloom. In summer, it is the turn of the white spires of the common spotted orchid (*Dactylorhiza fuchsii*) with a sprinkling of bee orchids (*Ophrys apifera*), twayblades (*Listera ovata*), and fragrant orchids (*Gymnadenia conopsea*). As well as orchids, the meadow is home to the rare meadow saffron (*Colchicum autumnale*), its naked purple flowers emerging in the autumn. This crocus is all that remains of the once extensive 'colchicum meadows', the remainder having either been ploughed or agriculturally 'improved'. A treatment for gout, colchicine is an alkaloid derived from meadow saffron, and has often been used to double the number of chromosomes in each plant cell in orchid breeding programmes. From the perspective of a commercial orchid grower, artificially producing tetraploids can be desirable, as these plants can be more robust, with larger flowers of thicker texture, for example.

Interest in creating flower-rich meadows has increased over the past few years in the UK, and gardeners are being encouraged to create their own mini-meadows as part of a drive to halt the country's alarming loss of biodiversity. The first step is to reduce the nutrient levels in the soil and to include yellow rattle (*Rhinanthus minor*) in the seed mix. Sometimes referred to as the 'meadow maker', yellow rattle is a semi-parasitic plant, that taps into the roots of grasses and reduces their vigour, depriving them of the ability to out-compete their neighbours.

In 1988, Dave Trudgill bought a small field in eastern Scotland with the intention of converting it into a wildflower meadow. The field was first treated with a herbicide, sown with a wildflower mix, including yellow rattle, but no legumes (which 'fix' atmospheric nitrogen and increase soil fertility). A pond was also dug to encourage aquatic life forms. After four years, the northern marsh orchid, *Dactylorhiza purpurella*, appeared on the pond spoil heap. A seed must have blown in on the wind. Locally sourced orchid seed was broadcast and nine different orchid species now flower on the meadow. Some are more successful than others.

In 2013, as part of the celebrations for the 60th anniversary of the Coronation of Queen Elizabeth II, Prince Charles (now King Charles III) launched the Coronation Meadows project, with the aim of establishing a new meadow in every county. A total of 90 new meadows, totalling over 400 hectares (1,000 acres), had been created by 2016. Orchids in different areas of the country may display different genetic signatures, and so in Worcestershire, the newly established meadow Far Starling Bank received hay from Eades Meadow, along with the seeds it contained.

The English 'Three Counties' of Worcestershire, Herefordshire, and Gloucestershire were once home to the largest and most diverse collection of orchards in the world, supporting an enviable variety of apples, pears, plums, and cherries. Up until the 1960s, seasonal workers from the 'Black Country' – England's industrial heartland in the West Midlands – would travel there to harvest the fruit. Sadly, the fate of the fruit trees mirrors that of the meadows, and around 85% have now been grubbed up as they have become uneconomic, and the countryside has lost the magical sight of thousands of trees in bloom in the springtime. Those that remain are managed by enthusiasts determined to preserve these precious remnants of days gone by. The grassland beneath the trees of a traditional orchard was usually managed by grazing livestock rather than machinery or chemicals and, if the grazing is carried out with the right intensity and timing, the grass beneath the trees would have a wildflower richness akin to a traditional hay meadow, with orchids such as the butterfly orchid (*Platanthera chlorantha*) and the pyramidal orchid (*Anacamptis pyramidalis*).

From maize field to orchid meadow
Samuel Sprunger has demonstrated that it is possible to convert what seem to be unpromising sites in the Swiss Jura into orchid-rich habitats. Two plots of agricultural land, which had previously either been improved grassland or used for the cultivation of maize or wheat, were purchased in 1998 and 2004 as part of an ecological restoration package in compensation for land taken for building a new road. The land had been fertilised with either manure and/or chemical fertilizers, and the crops sprayed with herbicides, fungicides, and insecticides.

The land faced south and was normally wet during the winter and spring, and dry in the summer and autumn. The soil was a thin layer of humus on top of a bed of mineral-rich soil on limestone. The plots were adjacent to pine woodlands on their southern margins, and hedges were either extended or new hedges with trees planted along the other field edges. The upper sections of the fields were stripped of their vegetation and soil. Both sections were sown with seeds of locally sourced wild flower mixes. A different seed mix was sown along the pine wood margin. The fields were rented out to a local farmer, with the stipulation that the grass must be cut for hay after June 15th. A second cut for hay, or grazing by cows, was allowed in September. The hedge and the fallow land were cut every second year. The use of fertiliser was prohibited.

In March 2004, Sprunger was amazed to find around 50 rosettes of *Ophrys apifera* and a few *Anacamptis pyramidalis* growing on the area where the humus had been removed, on the areas where it had been left, and in the hedge. The orchids were able to establish themselves where maize was being cultivated just six years previously! How was this possible? One possible explanation could be that the fungi living symbiotically with other native orchids in the nearby pine wood had colonised the soil of the lots. But how did the seed of the two orchid species arrive? The closest known location of populations of these two species were about 5 to 10 kilometres (3 to 6 miles) away and the seed was possibly carried on the wind from one, or a number of, these populations.

English flower-rich meadow in Worcestershire, UK, with *Dactylorhiza fuchsii*.
Photo: Joyce Seaton

247 SLIPPERS

Planting orchids in trees in Medellín, 1948.
Photo: EAFIT University

Five rosettes of *Anacamptis pyramidalis* and 153 rosettes of *Ophrys apifera* were found on the two lots during the winter of 2009–2010. Several other species of orchid had also appeared there. However, only *A. pyramidalis* was found where the site had been stripped of humus. Most of the plants bloomed in June and July of 2010, and most formed seed capsules. Based on observations of the development of the orchids, it was thought that it would be useful to delay mowing to maximize seed yield. As part of the ecological compensation for the construction of the motorway, different locations were linked by educational trails, which also join up with of the canton's hiking path network. An information leaflet is available for visitors to further promote interest in the area.

A roadside verge
Similarly, for 10 years Sprunger has been developing a new orchid habitat of five hectares (12 acres) in Hésingue, Alsace, from a derelict area beside a new road bordering an airport. After clearing the site of invading scrub and trees in 2013, the rough grassland was mown each year in August and September, and hay removed to keep nutrient levels low. The hay was dried, winnowed and used as fodder for farm animals. The first orchids appeared after five years. Today, more than 350 different species of flowering plants grow there, including seven orchid species. In the spring, the willows along the river banks produce pollen and nectar that the bees collect. In summer, thousands of insects forage on the many flowering plants.

Medellín, Colombia
A grainy black-and-white photograph published in a Medellín newspaper in 1948 reminds us that planting orchids in the branches of trees in cities is nothing new. The photograph shows a fire engine with the firemen climbing a ladder to plant orchids, probably cattleyas, in a kapok tree (*Ceiba pentandra*) in Avenida La Playa in front of the Junín Theatre and Hotel Europa. This was part of a plan by the city's public improvement society to beautify the urban environment. The buildings were demolished in the late 1960s. The tree is still there, but sadly the orchids are not.

When the photograph was taken, the population of Medellín, in the Aburrá Valley of the Department of Antioquia in the north-western cordillera of the Andes, was around 375,000. Today, Medellín and the surrounding metropolitan area is a megacity that is home to nearly 4 million souls. Such rapid urban expansion, together with high population density, brings many challenges, among them the problems of air pollution and the 'heat island' effect – when cities become significantly hotter than the surrounding rural areas – now exacerbated by global warming. Epiphytic orchids can act as indicators of ecological health, as 'canaries in the coalmine' or 'sentinel species', their variety and number declining in towns and cities where the atmosphere is drier and subject to traffic pollution. Not only are the orchids themselves sensitive to environmental stress, but so are their associated mycorrhizal fungi. In an effort to improve the environment for its citizens, Medellín has implemented the 'Green Corridors' project, planting thousands of trees and shrubs, thereby reducing the ambient temperature in parts of the city, and providing welcome havens of cool green shade. As well as inhaling carbon dioxide and exhaling oxygen, trees have the added benefit of acting as biological filters, their leaves removing health-damaging small particles from the air.

Visitors to Medellín can take a ride on the spectacular Metrocable (cable car) to visit Parque Arví, a reserve of around 16,000 hectares (40,000 acres) in the Central Cordillera, and one of the few protected areas in the Colombian Andes. Located 2,200–2,600 m (7,200–8,500 ft) above sea level, with daytime temperatures around 10°C (18°F) lower, it is distinctly cooler than in the city below. To restore degraded areas, conservationists need to know what the environment was like prior to the destruction of the forests. The forest remnants in the Aburrá valley have the potential to serve as a record of the species that remain in the region, but are no longer present in the vicinity of Medellín. The aim of the Arví project is to protect and enrich biological diversity in the area. To date, around 100 orchid species have been documented, both terrestrial and epiphytic. Conserving the other components of the environment is equally important. There are, for example, 12 species of *Anthurium* in the park, including the endemic and endangered black anthurium (*Anthurium caramantae*), and 19 bromeliad species. These are, in themselves, important indicators of the environmental health of the forest.

Orchids and bromeliads have been tied to many of the trees as part of the restoration initiative and are gradually becoming established. Many orchids can be found that bear seed capsules, indicating the presence of their pollinators, and the promise of natural dispersal of their dust-like seeds throughout the habitat. Towards the end of the pathway that runs through the park, visitors enter the Flower Dome. Constructed in the form of a geodesic dome, and covered with black wire netting, it replicates the cool, shady, and humid conditions of the surrounding forest. Here the tremendous diversity of the area's orchid flowers can be appreciated and photographed, from the tiny jewel-like *Lepanthes* to the much larger blooms of the many *Masdevallia* species.

Singapore: a garden without glass
'I like to think that some time in the future my grandchildren will look at orchids flowering in the trees and say: "Our grandfather planted those."' Tim Wing Yam, 2022

Founded in 1819 by Sir Stamford Raffles, today Singapore is a modern city-state. Originally, an estimated 82% of the island was covered with tropical lowland forest, but a combination of urban expansion, industry, and agriculture has led to less than 0.5% (192 hectares) of the former primary forest remaining today, much of it in the Bukit Timah Nature Reserve. Alfred Russel Wallace visited Singapore on several occasions between 1854 and 1862, at a time when the land was already being cleared. He tells us: 'There are always a few tigers roaming about Singapore, and they kill on an average a Chinaman every day, principally those who work in the gambir [*Uncaria gambir*] plantations, which are always made in newly-cleared jungle.' The last tiger in Singapore was shot in 1930.

With around 60 orchids remaining out of an original 224 species, and most of them endangered to a greater or lesser degree, in 1995 a recovery programme was initiated by Singapore Botanic Gardens with the aim of reintroducing plants into their natural habitats, parks, and roadside trees. Following a small-scale experimental laboratory micropropagation programme carried out between 1995 and 2008, the tiger orchid (*Grammatophyllum speciosum*), *Bulbophyllum vaginatum*, *B. membranaceum*, *Cymbidium finlaysonianum*, and *C. bicolor* subsp. *pubescens* were successfully reintroduced in and around Singapore.

Banded flower mantis (*Theopropus elegans*) has caught a Megachile bee while the latter tried to pollinate a tiger orchid (*Grammatophyllum speciosum*).
Photo: Tim Wing Yam

The programme has since been expanded to include a larger number of native orchids, focusing on critically endangered and vulnerable species. Between 2000 and 2015, 17 species were rediscovered. With their charisma, horticultural appeal, showy floral displays, and/or interesting vegetative characters, these species have been particularly valuable for educating and captivating the general public. Whenever possible, seed was obtained from wild-grown plants, because the embryos were the result of natural pollination and, therefore, were more genetically diverse than their counterparts grown in captivity – which had been selected over time for their most appealing and horticulturally desirable characteristics. Nationally extinct species have also been reintroduced and planted on roadside trees in parks and nature areas. To date more than 40,000 plants, representing 60 species, have been reintroduced in more than 40 different locations.

The rain tree, *Samanea saman*, was introduced into Singapore in 1876 from the tropical Americas. With its spreading, umbrella-shaped crown, it has proved to be an ideal street tree and, with its rough bark, makes an excellent host for epiphytic orchids. Older specimens that support other epiphytes tend to make the best hosts, often where water drains down into forks in main branches. Of course, the requirements of individual species will differ, and many orchids grow high up in the canopy. With judicious pruning over time, the density of the tree crown can be thinned and opened to allow in more light. Choice of a suitable microclimate is key. Seedlings in areas of relatively high humidity have higher survival rates than those transplanted into drier areas, and acclimatise better in the shade. The best time to reintroduce the orchids is during the wetter months of the year. Planting from October to November has proved to be most effective. Plants of the tiger orchid were found to survive better when planted nearer the ground in exposed regions, where the atmosphere was more humid and they were not subject to a lack of moisture. Remarkably, many epiphytic orchids survived a prolonged drought that occurred early in 2014, possibly because their large pseudobulbs stored water, and their thick, leathery leaves retained it. By contrast, many ferns, which lacked these water-saving features, perished.

Plants are regularly monitored to assess their progress. The long-term goal is that they become self-sustaining and maintenance-free. Seed capsules have begun to appear spontaneously, which suggests that some of the natural pollinators may still remain, despite the urban setting. The reintroduction of many species has been followed in many cases by natural regeneration. *Bulbophyllum vaginatum* was first reintroduced to Bukit Batok Nature Park in 2010, and in 2014, seedlings were found at the base of the reintroduced plants. Subsequently, seedlings were also observed on tree trunks where the species had been planted in other parks. Specimens of *Cymbidium finlaysonianum* were planted

In the early days, orchid seeds were germinated in conical flasks with cotton wool plugs, as can be seen in this grainy photograph taken at Singapore Botanical Gardens.
Photo: Philip Seaton

at Pasir Ris Park in 2008, and in May 2018, a healthy seedling was found on a tree growing opposite the reintroduced plants. By December 2020, the seedling had grown and a bird nest fern was growing at the same location. Another group of *Cymbidium finlaysonianum* was reintroduced to East Coast Parkway in 2009, and in 2015, an established and healthy seedling of the species was found on one of the trees where there had been a reintroduction.

Reintroduction of native orchids has led to the discovery of other components of Singapore's biodiversity. Forest-associated Megachile bees that visit *Grammatophyllum speciosum* in several sites have not previously been recorded in Singapore. *Pinalia bractescens* that was planted at Pasir Ris Park was pollinated by another bee, *Braunsapis clarihirta*. Natural fruit set has been observed in both instances. The reintroduced orchids attract many other insects, spiders, lizards, and even snakes. Future plans include the reintroduction of additional species, surveys of more remote areas within Singapore, and studies on the genetic variability of native populations.

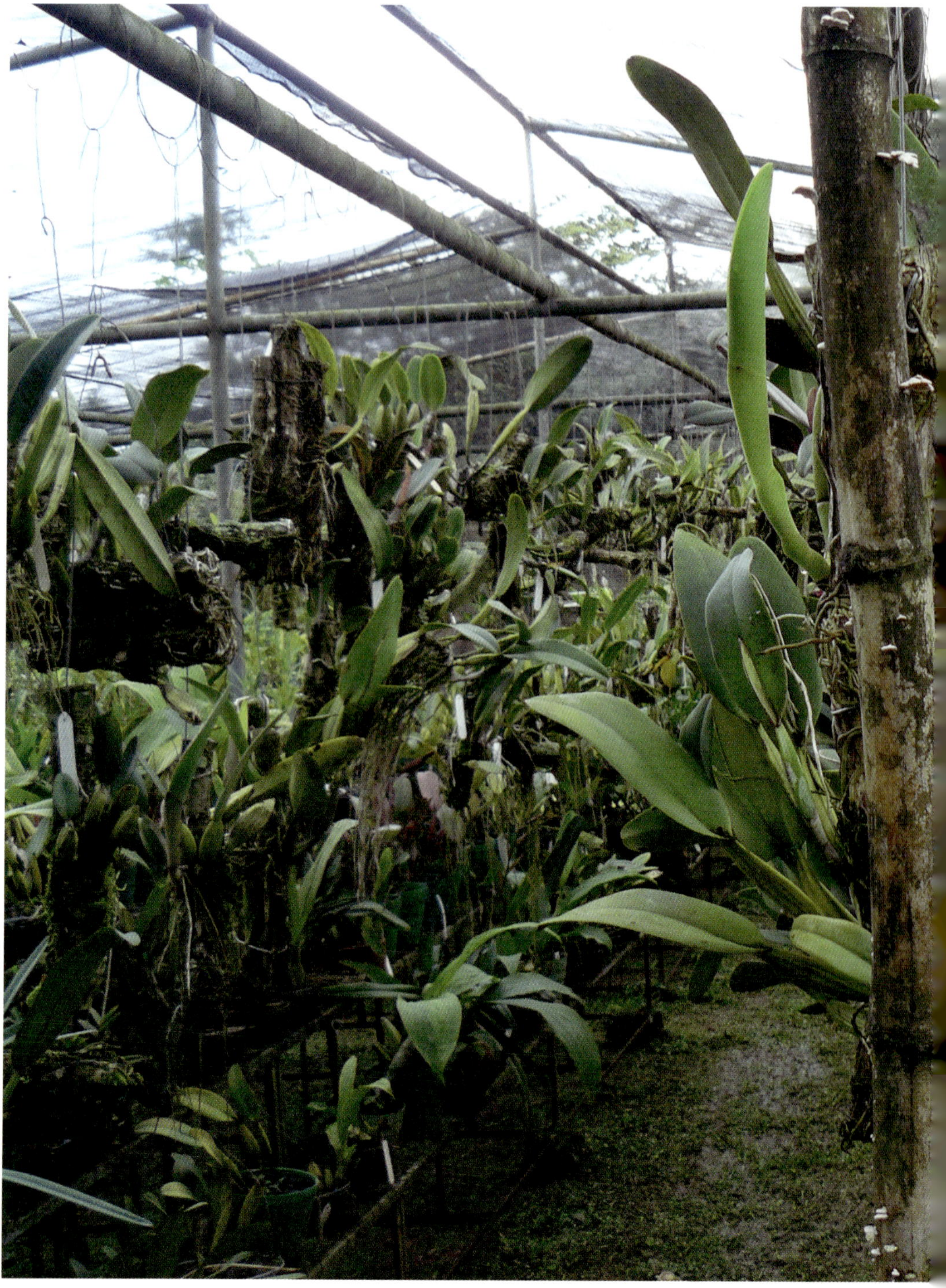

17 Living collections: safeguarding plants for the future

'John Lager, of Lager and Hurrell . . . had, about 50 clones [of Cattleya mossiae] that had been individually selected over the years from plants that he had collected in the wild . . . I often wonder if they are still alive today in various collections, but they have probably all been discarded in favour of modern hybrids.' Carl Withner, 1988

(previous pages)
Leticia Abdala of Vivero Media Dapa looking at a private collection of cattleyas in Colombia.
Photo: Philip Seaton

Entering a tropical orchid house on a frosty morning in temperate regions, you are transported to the rainforest by the breath of warm, humid air that steams up your glasses, and the delicious aroma of moist earth and hints of exotic perfumes. In the colder months, those of us living in cool temperate climes envy growers in the tropics. No need for glass. No need for expensive heating (or summer cooling). All that is required is some protection from the elements. Ideally, living collections should be maintained in their countries of origin. In the tropics, orchids are often grown in shade houses in a series of different environments (at different altitudes for example), each reflecting the provenance of the species in question. What, then, is the role of botanic gardens, especially where they are in temperate countries and hold collections of tropical orchids? In addition to their role as plant guardians, the answers lie in research, education, and the sharing of technical expertise.

Preservation of plants in living collections provides a safety net – some insurance against the extermination of species in the wild. The question is, do we know what species are in living collections, and do we know who holds these plants and where they are? Botanic gardens have become important centres for plant conservation, sometimes sheltering rare species. According to the Plant Conservation Report published by Botanic Gardens Conservation International in 2020, 38% of threatened orchid taxa are in *ex situ* collections. But does one individual plant count as an *ex situ* collection? The aim should be to conserve the maximum amount of genetic diversity for each species, to reflect as far as possible that found in the wild population(s). This can be achieved either within an individual collection, where a garden could hold populations of one or several different target species, or within a 'metacollection' – a network of collections that are spread across a number of centres.

To ensure that a species persists, plants should be maintained in duplicate collections in different locations wherever possible. Living collections require a long-term commitment to their maintenance if they are to make a significant contribution to conservation. They can be expensive to maintain, both in terms of staff costs and, in temperate countries, heating costs. The potential for heating failure is a constant concern in cooler climates, as is failure of cooling systems in hot climates. Living collections can be at risk from a combination of factors, including neglect, poor horticultural practices, and problems arising from pests and diseases.

Skilled gardeners and horticultural expertise are crucial to success when growing orchids for reintroduction, and it is important to maintain a pest-free environment. When visiting the *Cypripedium macranthos* collection in the mountains close to Beijing, Phil asked if the growers faced any particular pest problems. He was shown a large, pale beetle larva (grub) several centimetres long. This evil-looking creature is found in the soil, and chews through the bases of orchid shoots. But what was it? Phil was later introduced to Prof. Youqing Luo at Beijing Forestry University, an authority on wood-boring insects (his room is peppered with short logs of wood, standing on end around the walls, or used as stools for visitors, each with a small pile of sawdust at the base), who said it was the larva of a scarab beetle.

Nothing lives forever. An orchid collection is a dynamic entity, with the need to constantly regenerate/replace plants with ongoing propagation programmes. For example, both Kew and Atlanta Botanic Garden continually hand-pollinate their living collections. Selective breeding to retain (and enhance) the genetic diversity within collections is vital. There have been a number of proposals to set up orchid pollen exchange schemes. Information regarding the timing of pollen maturation on the parent plant, or how long the pollen remains viable *in situ*, is limited to a few species. Work with the common spotted orchid, *Dactylorhiza fuchsii*, at the Millennium Seed Bank, for example, indicates that it may be possible to store orchid pollen of at least some species for a number of years if it is dried and kept in an air-tight container in a refrigerator at 5°C (41°F).

Single examples of rare and endangered plants representing different genotypes (i.e., the plants are different clones) may be found in botanic gardens and other collections around the globe. Pollen exchange between collections has the potential to enhance collections by widening their genetic base through cross-pollination, thereby providing additional security for species in those collections. Seed that are produced in this way can be divided equally between participating institutes. Pollen and seed should be sent using express mail to minimise the impact of any potential adverse environmental conditions during transit.

With over 8.5 million items in the Millennium Seed Bank, including seeds representing 40,000 different species including orchid seeds, the Royal Botanic Gardens, Kew houses the largest and most diverse botanical and mycological collections in the world. They represent approximately 95% of vascular plant genera and 60% of fungal genera. The aim of Kew's Science Strategy 2021–2025 is to channel all its scientific resources and expertise 'to understand and protect plants and fungi for the well-being of people and the future of all life on Earth'. The Smithsonian Gardens Orchid Collection acts as the living collection for the USA and its territories. Established in 1976, it has grown from five plants to 6,000 plants from all over the world. It protects critically rare species under immediate threat, forming the basis for reintroduction and restoration programmes. The collection is particularly focused on Central America, the Caribbean, and

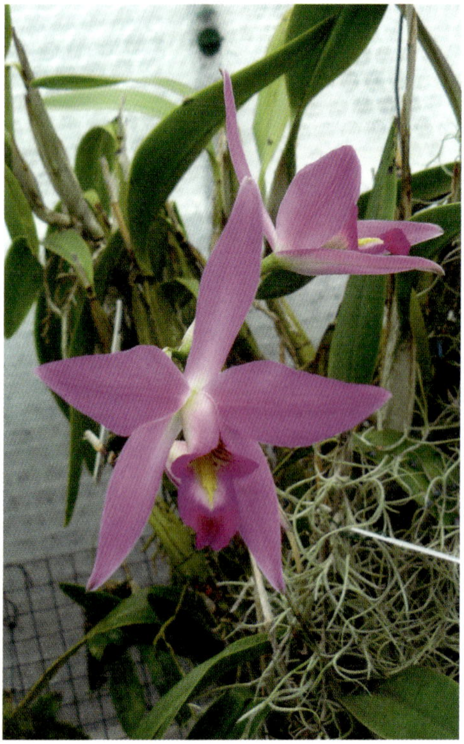

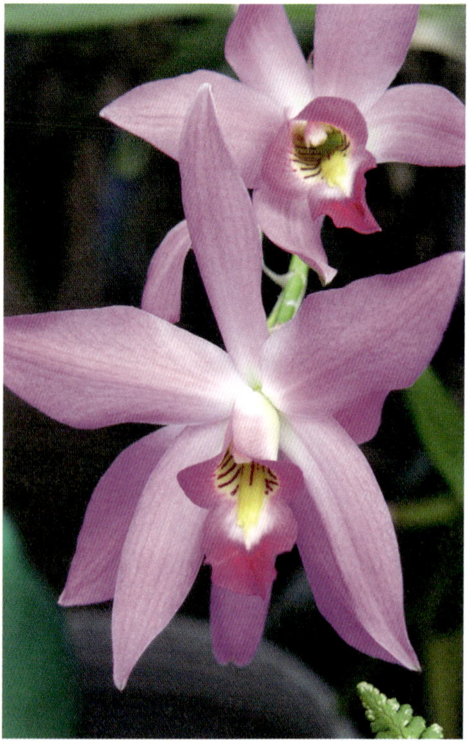

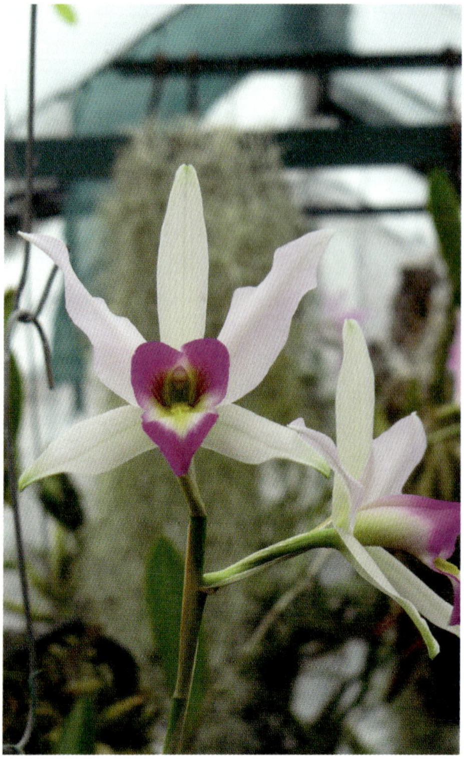

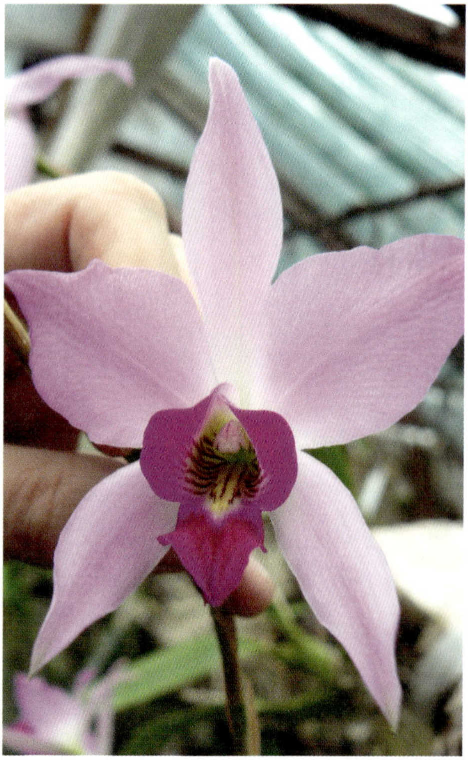

As an example of genetic diversity in one species, *Laelia anceps* exhibits considerable variation in its flowers.
Photo: Philip Seaton

South America, with a recent emphasis on endangered orchids and those of historical merit. The entire collection has been tested for odontoglossum ringspot and cymbidium mosaic viruses. Led by horticulturist Justin Kondrat, the collection looks to expand its outreach through promoting *ex situ* conservation. Unfortunately, botanic gardens in general often face problems with funding, and the retention of staff and their associated expertise (gardeners, in particular, are often undervalued and poorly paid).

One small botanic garden

In 1894, John Lager described *Cattleya quadricolor* growing in the swamps of the broad, flat valley of the Río Cauca in Colombia: 'Here this cattleya grows in forests, on level land to a great extent marshy and at times inundated, consequently the moisture the plants receive throughout the year is considerable. The evaporation of the stagnant water through the influence of the heat transforms it into a light mist which finds its way upwards among the trees and branches on which *Cattleya chocoensis* [*Cattleya quadricolor*] grows. The trees in this region are of a short and stunty growth, and they are mostly covered with decayed matter and vegetation of every description.' He continues: 'Here I saw the most beautiful sight it has been my fortune to see; in these jungles the plants grew by their thousands, the trees being literally covered with them and in full bloom. I particularly remember that I got my mule under a tree, and, sitting in the saddle, picked a large bunch of flowers; these were particularly fine and large, so different from what we generally see under culture, when they only half open.'

The trees have since been cut down, the swamps drained, and the land turned into one of the most fertile regions of Colombia. Today the valley is largely an immense monoculture of sugar cane, and the foothills of the bordering central and western Andean cordilleras are predominantly used for cattle grazing. Just 19% of the original tropical dry forest ecosystem persists, with a mere 2.6% of the remaining patches confined within protected areas. *Cattleya quadricolor* today is restricted to a few sites of tropical dry forest transitioning into montane forest along the valley margins. Because the flowers do not open fully in cultivation, remaining almost bell-shaped, and have a typically nodding habit, *Cattleya quadricolor* is not as popular as *C. trianae*. Nevertheless, it is a beautiful species, with a delicious perfume, and of considerable ornamental interest. Despite the extraction of wild plants being prohibited, there is ample evidence of ongoing removal of plants from natural populations. Safeguarding the species in a living botanical garden may therefore be necessary as a short-term solution to ensure its survival.

While the focus tends to be on large institutions, the many smaller and lesser-known gardens have an essential role to play, particularly in an orchids' country of origin. Take for instance

Jardín Botánico Juan María Céspedes, located in dry forest a little over 97 kilometres (60 miles) from Cali, in the district of Tuluá, Colombia. Here, *C. quadricolor* grows on the trunks and branches of large trees, typically caracolí (*Anacardium excelsum*), guácimo colorado (*Luehea seemannii*), and tachuelo berrugoso (*Zanthoxylum verrucosum*). Many are draped in enormous, ragged curtains of grey Spanish moss (*Tillandsia usneoides*) with *Cattleya quadricolor* occasionally nestling in the crooks of the branches. The seasonal dryness of the habitat is evidenced by the presence of an epiphytic cactus (*Rhipsalis*) sprawled along the trunks and branches of the trees alongside grey lichens and silver tufts of tillandsias.

Private and commercial collections
Although the huge commercial and private collections of the past – with old black-and-white photos sometimes showing thousands of representatives of individual species – have gone, there remain a number of private foundations around the world that grow large numbers of orchids. The Eric Young Orchid Foundation (EYOF) on the island of Jersey and the Mathers Foundation in the UK are two such examples. Although they focus on hybrids, both organisations aim to support and promote research and conservation, propagation and breeding of orchid species and their hybrids. EYOF provides an educational resource for schools on Jersey by offering a series of workshops designed for all age groups. Similarly, Vicente Perdomo and Leticia Abdala of Vivero Medio Dapa, a commercial nursery on the edge of Cali, Colombia, host courses for amateur enthusiasts.

   Commercial growers grow good plants. Of course they do, their business depends on it. Amateur growers often focus on the 'best' forms that don't necessarily reflect the 'typical' species. On the other hand, commercial nurseries, by growing from seed and breeding 'superior' varieties at affordable prices, can reduce the pressure on wild populations. Hybrids are usually easier to grow and their widespread availability can reduce the appetite for wild-collected plants. Commercial nurseries may also hold important plant collections that may include old species (and hybrids). Certainly, some orchids can be very long-lived, and will persist in cultivation almost indefinitely if well-grown. Remarkably, some of Joseph Charlesworth's original plants, such as *Odontoglossum crispum* 'Avalanche' may still be found in collections as 'heritage orchids'. In the USA, Arthur Chadwick wrote in 2003 that his nursery still had examples of *Cattleya* species that were collected in the 19th century. *Angraecum longicalcar* is now rare in the wild in Madagascar (see 'All (dusty) roads lead to the lab', page 79). Marcel Lecoufle, of Boissy-Saint-Léger, France, who built up a collection of Madagascan orchids from the 1930s onwards, received seeds of *A. longicalcar* in 1962. According to Lecoufle, 'A picture was taken

Cattleya quadricolor with seed capsules at Jardín Botánico Juan María Cespedes, Tuluá, Colombia.
Photo: Philip Seaton

[presumably of a plant in flower] by Doctor Jean-Pierre Peyrot in 1962 near the Itasy lake.' Two months later, Peyrot found a 'black powder' coming from the fruits which he sent Lecoufle in an envelope. Since that time the species has continued to be available from seed.

All too often the word 'amateur' is associated with the word 'only', as in 'I am 'only an amateur.' This is a mistake. Amateurs (hobbyists) often have considerable expertise, particularly when it comes to the cultivation of their plants, and they often have time to devote to the care of more fastidious species. These people represent a small but important group of dedicated individuals who are often passionate about one particular type of orchid. For some, it is the members of the sub-tribe Pleurothallidinae that floats their boat, while for others, it may be a particular genus. Some maintain important living collections, such as the UK's orchid

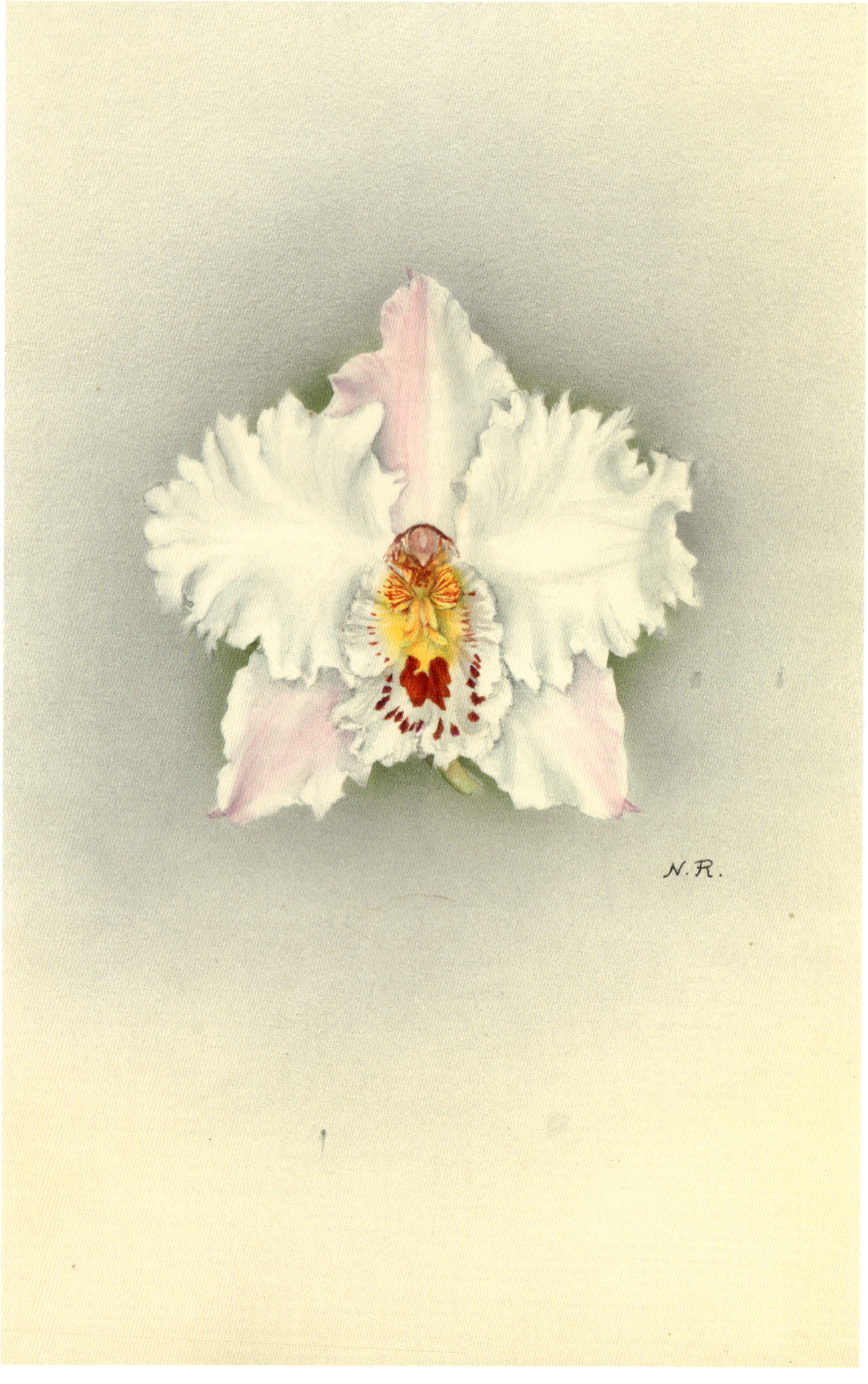
N.R.

Odontoglossum crispum 'Avalanche' was awarded a First Class Certificate by the RHS on 8 April, 1931.
Painting: RHS Lindley Collections

National Plant Collections, which act as a reservoir of endangered plants and their varieties. Sometimes, and often unbeknown to the grower, they hold important, or unique specimens. Such plants, when sufficiently large, can be divided and shared according to the gardeners' adage: 'If you want to keep a plant, give a piece away.' Orchid society members have donated the seed of plants that are now rare in cultivation, such as *Paphiopedilum haynaldianum* var. *album* and *Mexipedium xerophyticum*. The resulting seedlings can be shared with other growers and botanical gardens, thereby enhancing living collections. Some enthusiasts have their own home laboratories, where they raise the seedlings of species that are becoming increasingly rare in cultivation.

A National Plant Collection can also serve as an educational tool, increasing awareness among orchid growers and the general public of the enormous diversity of the orchid family. Individual collections can make an invaluable contribution to orchid science by providing seed for storage and scientific research and by maintaining important records in the form of photographs and herbarium specimens.

A research bacteriologist by profession, Michael McIllmurray took up orchid growing as a retirement hobby in 1993. He soon decided that he wanted to turn his pastime into something more ambitious, to make a more enduring contribution to botany. In 2001, his collection officially became the foundation of the UK's National Plant Collection of *Maxillaria*, with around 500 plants representing roughly 275 species. Building and maintaining a National Plant Collection takes a special set of skills, passion, and determination, bordering on obsession. It also requires dedication, curiosity combined with an attention to detail, and someone who is organised and able to maintain meticulous records. Aside from the value of the living plant collection itself, McIllmurray's collection was underpinned by a series of volumes of drawings and notes on each species. Each plant was photographed when in flower, and there was a comprehensive slide collection. Some plants were represented as paintings. Kew provided herbarium sheets for pressed specimens and there was a corresponding spirit collection of flowers. When McIllmurray sadly died a few years ago, his collection was passed on to Kew, and will remain a lasting testament to his commitment to orchid conservation.

Orchid viruses

'Fear not the orchid virus nor the bug. Instead, abhor the grubby hand that holds the unsanitized cutting tool and act accordingly.'
F. William Zettler

As important as living collections may be, keeping orchids healthy in captivity under one roof has its challenges. Viruses illustrate this point well. They are everywhere – in water, on land and in the air, but we just can't see them. Upon entering an unsuspecting organism, they quickly replicate themselves by hijacking a host cell's genetic machinery for their own dastardly purpose; they then look for a way out of the cell and into the world once again where they can re-infect new hosts *ad infinitum*. To do so, most plant viruses need tiny sap-sucking insect vectors, such as aphids or thrips. Paradoxically, the two most prevalent viruses infecting orchids, namely cymbidium mosaic virus (CyMV) and odontoglossum ringspot virus (ORSV), rely on humans to be transmitted.

Both are very, very stable, meaning that they can remain in an infective stage outside of the orchid for long periods of time. They are also capable of infecting all major orchid groups and, therefore, no orchid is immune. Once they gain entry, a seemingly healthy orchid may be infectious because symptoms are not always readily apparent. How do they gain entry? Neither virus is wind-borne nor transmitted from parent to offspring through seed under natural conditions. Thus, CyMV and ORSV need a little help – our help – to spread. Essentially, we provide these viruses with a free pass and a dream vacation when we cultivate them for long periods of time.

Cultivated orchids get sick from these viruses wherever we grow them. The unifying mode of transmission is through cuttings made using sharp instruments that are contaminated with virus-laced plant sap. The early collectors may have unknowingly initiated the spread of viruses as they loaded hundreds of exotic orchids onto ships, to be showcased in captivity under glass in faraway lands. When this contamination occurred remains unclear, but the spread of viruses may have accelerated once an especially appealing orchid had been selected for widespread cultivation, which would have been 1821 when the first commercial nursery, Loddiges, was established near London, England. This timeline also agrees with a colour plate published in 1900 depicting a *Cattleya* hybrid displaying symptoms typical of ORSV infection. One by one, cutting by cutting, these showy orchids became infected by the instruments used to multiply their numbers. Eventually, many of the orchids in our living collections became infected, which is where we find ourselves today. The global transmission and

> Meristemming is used commercially to produce vast numbers of genetically identical orchid hybrids (clones) for the horticultural industry. The meristem is the tiny ball of rapidly dividing cells at the tip of a plant's growing shoot. (Strictly speaking, what is being cultured is the single meristematic cell at the shoot's apex along with a small number of daughter cells). This technique also has the potential to help conserve rare orchid varieties when only one or a small number of plants remain. Orchids can live for many years in cultivation, and divisions of some of the original importations probably still exist in collections, both in their native countries and abroad, and may contain valuable genetic material/diversity. Plants that have been in cultivation for many years may gradually begin to deteriorate and may become infected with viruses. The meristem remains virus-free, however, and so newly invigorated, virus-free plants can be produced from it. Once the meristem has been obtained, the art is to persuade this minute ball of cells to continue dividing and to differentiate into a plantlet on an agar medium in a test-tube.

Francisco Merchán and Paola Jaramilla demonstrating the meristemming of an orchid to students at Jardín Botánico de Quito, Ecuador.
Photo: Philip Seaton

spread of these two viruses has occurred within the past 50 years, coinciding with the large-scale trade in orchids propagated *ex situ*. Given the strict regulations that limit the global movement of material, the spread of CyMV and ORSV should not have occurred so rapidly. Nevertheless, these viruses now occur practically everywhere where humans grow orchids.

The longer that wild, virus-free orchids are in cultivation, the more likely it is that they will become infected because of the frequency with which they are handled. The risk for virus transmission is also greater when many orchids are clustered together in high densities. What steps should orchid conservationists take? Considering that orchids do not pass viruses to their offspring, the most obvious path forward seems to favour artificial seed propagation, followed by release of genetically diverse seedlings back into suitable habitats as soon as possible, while still maintaining reasonably high survivorship.

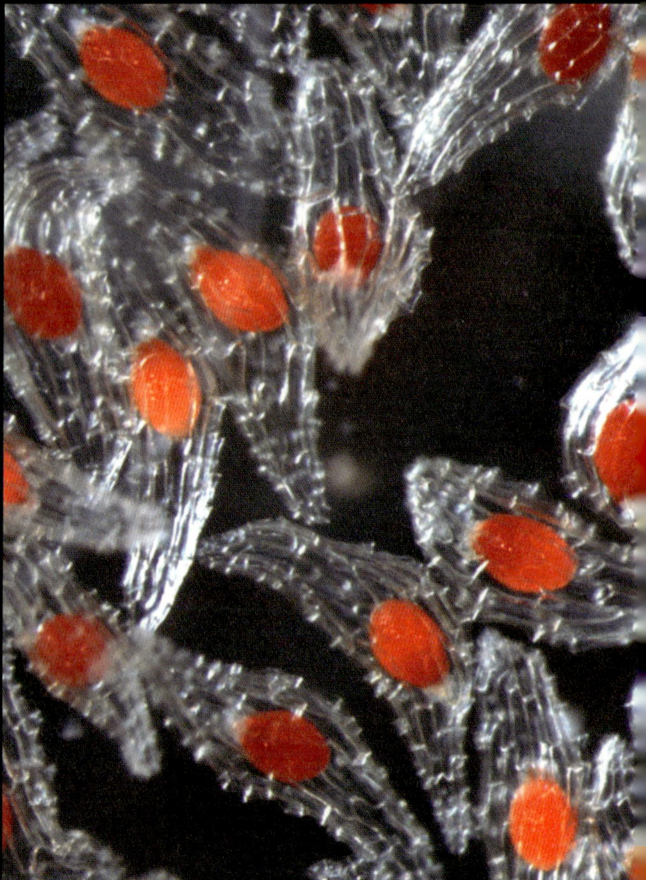

18 Seed storage

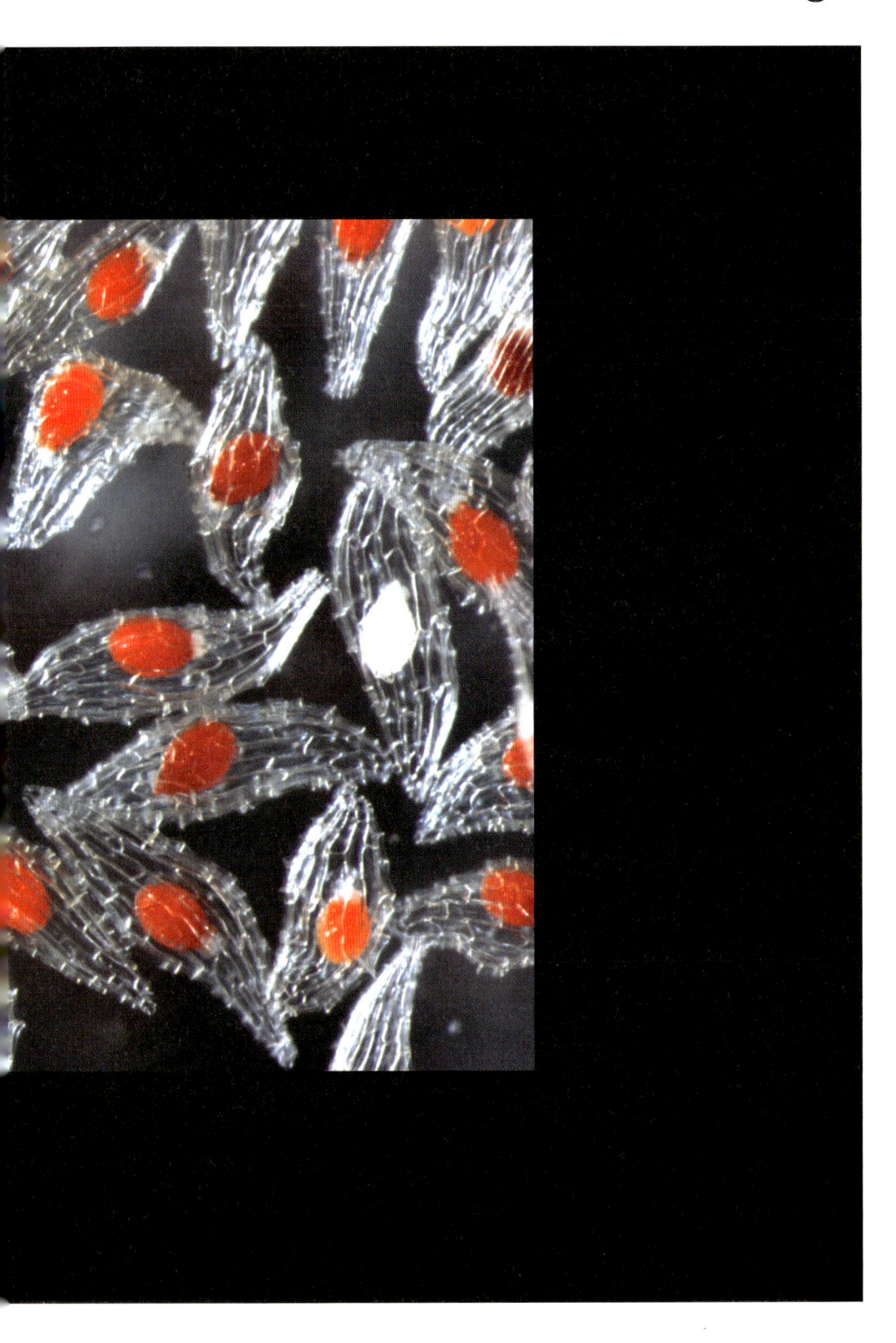

'Seeds are time capsules, vessels travelling through time and space. In the right place at the right time each seed gives rise to a new plant.'
Rob Kesseler and Wolfgang Stuppy, 2006

We owe an enormous debt of gratitude to Nicolai Vavilov – the first person to fully appreciate the importance of preserving plant genetic diversity. In the early 20th century, he visited five continents, collecting seeds of the wild relatives of crops, and establishing their centres of diversity. During the 900-day siege of Leningrad (today's St Petersburg) in the Second World War, the Russian scientists responsible for what was the world's first seed bank guarded Vavilov's collection of precious agricultural seeds. Nine eventually died of starvation rather than consume the precious crop seeds, and the unique genetic diversity within them. Although it seems unlikely that scientists would sacrifice their lives for orchid seeds, conserving their genetic diversity is nevertheless important.

Thirty years ago, however, there was little interest in storing orchid seeds, which were considered to be exceedingly short-lived. Once botanists had discovered that it was possible to store orchid seed, often for decades, attitudes changed. This realisation occurred at a critical time, because seed storage may be a last resort for saving some endangered populations – an insurance policy against future losses. Because they are so tiny, large numbers of orchid seeds occupy a very small volume, and seed banking is therefore an efficient and cost-effective way of preserving a large amount of a species' genetic diversity. Seeds representing all of the planet's orchids could be stored in a space no larger than that occupied by three domestic freezers.

The Millennium Seed Bank
Located in the English countryside south-east of London, Kew's Millennium Seed Bank opened in August 2000. It is the 'Rolls Royce' of seed storage facilities. The vault is a bomb-proof underground bunker, with half-metre-thick concrete walls, designed to last for 500 years. Safe from dust and monitored for radioactivity by a sensor in the roof, emergency pumps are kept on permanent standby to guard against flooding. After descending a gleaming stainless-steel spiral staircase, access to the seed bank is through a heavy, canary-yellow door. On opening the door visitors find themselves in a small atrium. After closing the outer door, they are able to open a second door to enter the vault's spacious preparation room, which is maintained at 15°C (59°F) and 15% relative humidity.

Along one wall of the preparation room are doors providing access to cold rooms, which are maintained at −20°C (−4°F). To be allowed to enter one of the cold rooms, the visitor must first fight his or her way into a royal-blue, fur-lined all-in-one. Fur boots,

(previous pages)
Orchid seeds stained with TZ reagent. Embryos that are red indicate that the seeds are viable. The single (white) unstained embryo indicates that that seed is not viable.
Photo: Jonathan Kendon

(above)
Visitors are able to watch scientists at work at the Millennium Seed Bank, Royal Botanic Gardens, Kew.
Photo: Philip Seaton

(right)
Scanning electron micrograph of an orchid seed of the eastern prairie fringed orchid, *Platanthera leucophaea* from North America.
Scale bar = 200 µm.
Photo: Kevin Gribbins

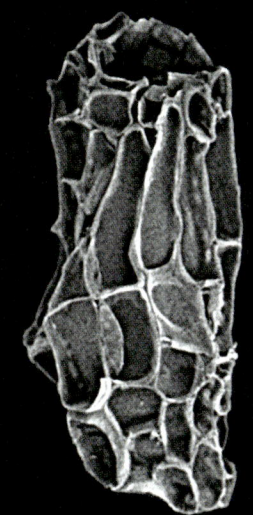

fur mittens, and a peaked hat with earmuffs, complete an outfit that wouldn't look out of place on an Arctic expedition. Each cold room has an alarm system, that is set for a maximum of 20 minutes. The walls are lined with banks of shelves containing glass Kilner jars, full of seeds of different plant families in all sizes, shapes, and colours. Smaller seeds are stored in smaller glass vessels in banks of drawers. In this high-tech world it is easy to forget that low-tech solutions are still sometimes the best. Found in kitchens around the UK, Kilner jars turn out to be the perfect jar for seed storage. The combination of a natural rubber ring and a metal clamp provide a perfect seal, allowing seeds to be stored for many years at their optimum moisture content. Orchid seeds, by contrast, are stored in small, hermetically sealed glass vials or foil laminate bags. The glass vials are frequently kept inside Kilner jars together with a sachet of self-indicating silica orange, to confirm that the atmosphere within remains dry (silica orange, gradually turns green when it absorbs moisture in a humid atmosphere). Each collection is divided into two samples: one for long-term storage, and the other to be routinely accessed for germination to ensure continued viability.

Life in the freezer

How long can you store orchid seed? It depends on a number of factors. First, the initial 'quality' of the seed – the percentage of viable seeds. It depends on how the seed has been stored. Has it been dried to the 'optimal' seed moisture content? Has it been stored at a low (refrigerator or freezer) temperature? Under identical conditions, as is to be expected, seed of individual species will remain viable for different lengths of time. What we can say with confidence is that the seed of many species will survive for many years, the initial percentage germination is high, it has been dried to a suitable (low) moisture content, and it has been stored at a low temperature in a refrigerator or a freezer. Nelson Machado-Neto and his co-workers, who stored the seeds of eight Brazilian *Cattleya* species as part of the OSSSU project (see 'Orchid Seed Stores for Sustainable Use', page 276), recently found that six species were at least moderately long-lived (with half-lives of around 30 years) and two were long-lived (with half-lives of more than 30 years – half-life being the time it takes for the percentage germination to be half of the initial percentage germination). Critically, they found that post-storage treatment was key to success in germinating the seeds. For optimum germination, seeds that have been stored in a refrigerator or freezer should be equilibrated at room temperature for 24 hours or in 10% sucrose solution before sowing. As small as these details may seem, unlocking the fastidious germination requirements of orchid seeds is of paramount importance to conservation.

Over time, the number of seeds in any seed lot that retain the ability to germinate declines. Because of the natural genetic variability within any seed lot, a small number of seeds will have comparatively short lives, others will have long lives, with the majority somewhere in between. Measuring the percentage germination of stored seeds at intervals allows you to plot a graph and, using the magic of mathematics (plotting percentage germination on a probability scale) – you get a straight line. As a result, it is possible to predict the future percentage germination of the seed lot of that species over many years.

Enlisting young people
'Somewhere, something incredible is waiting to be known.'
Carl Sagan, 1977

Saving orchids cannot be successful unless we enlist the help of more young people, but in today's modern technological world, this seems easier said than done. Orchid societies have a lot to offer, but it is often challenging to find members at monthly meetings who have yet to sprout a single grey hair. There is no shortage of young people eager to get involved, and for many years, the American Orchid Society has connected young people to orchids through grants and awards aimed at college students, for example. The mission of the society is to promote and support a passion for orchids through education, conservation, and research. In 1989, Lawrence Zettler was one such recipient as a new graduate student at Clemson University in South Carolina. Using society funds raised from the sale of Marion Sheehan's colourful artwork, he discovered a new fungus species and used it to successfully germinate seeds of an endangered terrestrial orchid (*Platanthera integrilabia*). A career in conservation was born.

In 2016, Rachel Helmich carried out an experiment at Illinois College in Jacksonville that raised more than a few eyebrows. As a college student in her early 20s, she asked many questions that often started with, 'Why?' or 'Why not?' One day, she asked, 'Why don't we try to germinate orchid seeds and fungi that have been stored in the refrigerator for 20 years?' Her professor replied, 'Because the seeds are probably dead by now and the fungus is too old.' Undeterred, she followed with, 'Then let me at least try.' Reluctantly, the professor responded, 'Well then go ahead, but you're wasting your time.'

After reaching into the lab freezer, she removed seeds of *Platanthera integrilabia*, wiping off ice crystals from the exterior glass tube before opening its contents which had been sealed for 28 years. *Platanthera intergrilabia* is now listed as a US Federally listed species, so germinating these seeds had added meaning. Reaching into the refrigerator below, she then pulled out two cultures of a mycorrhizal fungus (*Tulasnella inquilina*) that had germinated those same seeds. After inoculating the seeds with the fungus on oatmeal agar within Petri dishes, the experiment officially began.

The dishes were allowed to incubate at room temperature in a neglected corner of the lab. After a couple of months, Helmich could not resist the urge to check them. After peering into one, she spotted healthy robust protocorms from seeds collected at a time before she was born. The protocorms eventually formed leaves and grew into healthy seedlings suitable for reintroduction back into nature. To say she was pleased would be an understatement.

Helmich's experiment not only demonstrated that the seeds and the fungus had remained viable, the percentage of seeds that germinated and developed to the leaf-bearing stage closely paralleled the results from the original experiments carried out 28 years prior. Her reward for hard work and perseverance culminated in a research poster that she presented at the 22nd World Orchid Conference in Guayaquil, Ecuador, in 2017. Today, she is the micropropagation specialist at the Missouri Botanical Garden. Sometimes elders need to be schooled.

(above left)
Flowers of *Platanthera integrilabia* shown during peak flowering in August 2022, in the Daniel Boone National Forest, Kentucky, USA. This threatened terrestrial orchid appears to be pollinated by both butterflies and moths.
Photo: Lawrence Zettler

(above right)
Hand-pollination of *Cattleya amethystoglossa* in the shade house of Nelson Machado-Neto. The anther cap can be seen on the tip of the cocktail stick.
Photo: Philip Seaton

Plant identification

First and foremost, when storing seed, you need to be sure that the species in question is what you think it is. Suppose a species has been found to be extinct in the wild. Fortunately, you have seed of that species in your seed bank. You sow the seed. It germinates. You continue to grow the seedlings in flasks until they are large enough to transplant into compost. You transplant them. You grow the seedlings for several years until, eventually, they flower . . . and it is a different species! From the perspective of orchid seed banking and conservation, such errors are disastrous, and must be avoided at all costs. Robust identification of species is vital. Voucher specimens are essential, either in the form of dried herbarium specimens or, where this is not practical (for example, where the plant is rare, or there are only a small number of plants in the population), photographs can be taken instead.

Harvesting seed

The life of a seed collector is not an easy one. Harvesting seed from distant populations can be challenging and expensive. Populations may be widely distributed, difficult to access, and involve travelling to the site(s) multiple times.

Let's assume that it's your lucky day. The weather is dry (unless you are harvesting seed in a cloud forest of course), slugs or snails or other wildlife haven't eaten the capsules, neither are there any insect larvae burrowing into them. Moisture is your enemy because not only can it lead to the growth of moulds, making it virtually impossible to surface-sterilise the seed, but a high seed moisture content also considerably reduces the lifespan of the seed's embryo (yes, transferring seeds to polythene bags is bad news). Seed must then be separated from debris such as frass (insect poo) and from the capsule material to reduce the risk of contamination. Seed should then be dried as soon as possible. Consequently, it is sometimes easier to use plants in living collections as a source of seed, at the same time being aware that such seed collections may have reduced genetic diversity when compared to seed derived from a wild population. In this sense, seeds derived from wild populations are genetically superior.

The skill and experience of gardeners is important. Healthy, well-grown plants produce larger capsules containing greater numbers of seeds. Plants can be pollinated by hand. Flowers of some species remain open for a single day, whereas the flowers of others remain open for two months or longer. To be able to store the maximum amount of genetic diversity, out-crossing between two genetically distinct individuals is preferred. Where only a single plant of a rare species is available, however, self-pollination may be the only option. A few species have unisexual flowers, and some are self-incompatible, adding another layer of complexity.

When harvesting seed, timing is everything. Imagine, after

weeks or sometimes months of careful observation, walking into the greenhouse one morning to find everything covered in a fine 'dust'. The long-awaited seed capsule has split overnight, and the gentle breeze from the air-circulating fans has dispersed its contents throughout the greenhouse. The aim is to harvest mature seed, just as the capsule begins to split, and so daily monitoring of capsules towards the end of the ripening period is vital. Capsules can be enclosed in breathable bags – tea bags, with their multiple perforations, are ideal for enclosing small capsules. Immature seed, collected from an unripe capsule – the so-called 'green pod' – is often used in commercial laboratories, but is not recommended for seed banking because their embryos generally have poor survival in storage.

Cleaning, drying, and storing

Once harvested, drying the seed to a suitable moisture content as soon as possible is critical to its future survival in storage. Grandma used to put a little rice in the salt cellar to absorb any moisture and keep the salt flowing freely. Similarly, the amateur grower can use dried rice – heated in an oven at 80–100°C (176–212°F) until it appears to be slightly toasted – to dry orchid seeds. Once again, Kilner jars, with their air-tight lids, make ideal containers for drying seed.

As you would expect, scientists use slightly more sophisticated methods. When seeds arrive at the Millennium Seed Bank, they are taken to the drying room, where they remain for a minimum of seven days at 15°C (59°F) and 15% relative humidity, allowing them to equilibrate to the required moisture content. Not every institute has a drying room, however, and seed can also be dried in a glass desiccator over a saturated solution of either calcium chloride or lithium chloride, which produces the ideal seed moisture content.

Visitors to the Millennium Seed Bank or Atlanta Botanical Garden (ABG) can view scientists working in their labs through plate glass windows. On stepping over the threshold of the lab at ABG, before donning a crisp, white lab coat, visitors are instructed to take off their shoes and select a pair of freshly cleaned 'crocs' for walking over a spotlessly clean tiled floor. Avoiding contamination of flasks is vital, and everything must be kept spotlessly clean and dust-free. Success depends on practising good sterile technique. Today's technicians carry out their seed sowing and transfer of seedlings to fresh medium in laminar flow hoods with HEPA (high-efficiency particulate-absorbing) filters that

> An invasion of dust mites (*Dermatophagoides* – literally 'skin eater' – species) is every lab technician's worst nightmare. Able to squeeze through the tiniest of cracks, the mites are too small to be seen with the naked eye, along with the millions of bacteria and fungal spores that cling to their bodies and all eight of their tiny legs. Their presence is revealed by tell tale trails of gooey bacterial colonies meandering across the surface of the medium in flasks and Petri dishes, marking where the mite once crawled days or weeks previously. By the time the damage is discovered, the mite has crawled off to another dish or flask.

Jason Ligon and Yanny Vasquez working in the micropropagation lab at Atlanta Botanical Garden.
Photo: William (Grant) Morton

remove bacterial and fungal spores, and provide a clean laminar flow of air over the area where sterile work is being conducted. Brilliant if you suffer from hay fever! Nevertheless, with care, it is possible for today's amateur to achieve similar results using a simple glove box.

Germination and viability testing

There is no point in storing dead seed. Once harvested, you want to know if your seed is viable or not. Examining seed under a microscope will tell you whether the seeds contain embryos. Sometimes a seed capsule will contain nothing but 'fluff' – they are seeds without embryos. Assuming that there are seeds with embryos in your sample, you then want to know if the seed will germinate. Everyone has their favourite way of sowing seeds. At

Kew they favour the 'packet method', putting the seed in little envelopes made of unbleached coffee filter paper that are then sterilised in a disinfectant solution. The packets are then washed in sterile water, before being opened, and the seed blotted on the surface of the culture medium. At Atlanta Botanical Garden, the seeds are sterilised in a test-tube of disinfectant solution, then the seed suspension is filtered through a sterile filter paper. Finally, the seed is sown by dabbing the filter paper on the surface of the culture medium. Both methods work equally well, as do many other techniques. Seed is sown on an appropriate germination medium and germinating versus non-germinating seeds are counted. The art is not to sow too many seeds at a time, because when crowded together, they are very difficult to count.

Unfortunately, standard germination tests often take weeks, sometimes months. One alternative is to use a chemical stain – 2,3,5 triphenyl tetrazolium chloride (TZ) – to determine viability. Embryos that are alive (viable) are rendered red whereas embryos composed of dead tissues do not change colour. Unfortunately, not all seeds respond well to TZ testing, especially where the seed coat is dark in colour (often a problem with terrestrial species). Sometimes it fails to yield clear results. Is that red, or pink? But when it does work, clear results can be obtained in a couple of days.

Subsequent regular testing is essential for monitoring the quality of stored seed. This provides an indication as to when accessions should be renewed. Although the importance of getting good quality seed cannot be over-emphasised, accessions containing a low percentage of viable seeds can still be valuable, especially where the species in question is of high conservation importance.

Orchid Seed Stores for Sustainable Use
For many years, one of us (Phil) and Hugh Pritchard, at the time Head of Seed Conservation Rearch at the Millennium Seed Bank, had dreamed of setting up a global network of orchid seed banks. In 2007 their dream came true. Kew was awarded funding for a three-year UK Darwin Initiative project, Orchid Seed Stores for Sustainable Use (OSSSU), led by Hugh, with Phil as project manager. Universities and botanical gardens from 15 countries – seven in Asia (China, India, Indonesia, Philippines, Singapore, Thailand, Vietnam), and eight in Latin America (Bolivia, Brazil, Chile, Colombia, Costa Rica, Cuba, Ecuador, Guatemala) – participated in the project. Workshops took place in the autumn of 2007 in Chengdu, China, and at Quito Botanical Garden, Ecuador, to enable participants to share expertise and establish common protocols. A final workshop also included representatives from the US, Panama, the Dominican Republic and Estonia, and was hosted at the University of Costa Rica in 2010 to exchange findings and record data.

The large size of the orchid family means that we only have data for a comparatively small proportion of the total number of species. There remains a considerable amount of work to be done to identify those species that have unusually short-lived seeds, and to consider possible solutions. The answer may lie in storing seed in freezers maintained at –80°C (–112°F) or –140°C (–220°F), or in liquid nitrogen at –196°C (–320°F), each of which may require different seed moisture contents. If samples of seed were deliberately set aside for testing 50 years from now (a time capsule), future generations of seed technologists would be delighted.

The aim of the OSSSU project was to generate data on seed germination and viability, and seed capsule ripening times, which is important to know when harvesting seed. One portion of seed was sown on Knudson C medium, enabling comparisons of the germination of a wide range of species and genera, originating in different areas of the world, on one common medium. A second portion of seed was sown on whatever the participants in their experience found to be the 'best' medium. The aim was to test the germination and viability of selected species at regular intervals over the period of the project and beyond. Seed-raised plants that were produced as a result of germination tests provided material for both reintroductions and the enhancement of existing populations, or to enhance living collections.

What next?

What proportion of the world's orchids are currently secured in seed banks around the world? The Millennium Seed Bank currently stores orchid seed representing around 1,370 named species – somewhere between 4.5 and 5.0% of the world's orchid species. In China, the Kunming Institute of Botany seed bank currently has 415 species in storage, representing around 27% of the country's orchid flora. In 2017, Kingsley Dixon and Nigel Swarts reported that the Orchid Seed Bank Challenge, supported by the Millennium Seed Bank Partnership, had stored three quarters of the 408 native terrestrial orchid species of the south-western Australian global biodiversity hotspot, and was well on its way to achieving 100%. The Seed for Life project had collected at least 25% of Kenyan orchid species.

This is an encouraging start, but clearly a more extensive global seed bank network is needed, with each country storing seed representing its own orchid flora. OSSSU didn't end with the workshop in Costa Rica, however. Funding was obtained to deliver a workshop at the Millennium Seed Bank for holders of UK National Orchid Collections, and a workshop for partners in Latin America was held at Jardín Orquideario de Soroa in Cuba in 2012. Interest in seed banking as a conservation tool continues to grow. For example, Vincent Droissart reports that more than

Orchid seeds at Kew's Millennium Seed Bank are stored at −20°C in an underground vault.
Photo: Philip Seaton

300 orchid species have been conserved in freezers in local seed banks in Yaoundé in Cameroon and Antananarivo in Madagascar. New seed banks are being developed in countries as far apart as Bhutan, Sri Lanka, and Bolivia. Further training workshops have been delivered in additional countries in Europe, Asia, and Latin America. The aim is to teach, train and, above all, inspire people of all ages. Passing on expertise through cascade training of the next generation is vital. Kanchit Thammasiri at Mahidol University in Bangkok, Thailand, for example, has taught conservation courses focusing on maintaining genetic diversity, practical methods, breeding, and sustainable use to people from a range of

backgrounds and professions, including students from schools and universities, teachers, lecturers, growers, merchants, engineers, and medical doctors. Meanwhile, regular testing continues in many centres. Seeds are removed from storage every five years and tested for viability at the University of Cuenca in Ecuador, for example.

Orchids don't respect the geographical boundaries between countries, and seed from the same species may be stored in the seed banks of neighbouring countries. In a small country like the UK, with around 50 species, it is possible to store them all. The North American Orchid Conservation Center focuses on the 200 or so species native to the United States and Canada. But, what about Colombia, with 4,270 species? Storing seed samples representing all of the country's species is a daunting commitment, and it may be wise to prioritise those species that are currently most vulnerable to extinction. Storing seed in duplicate institutes for security purposes is good seed banking practice, and ideally each country would store duplicate samples in different locations within that country. If they so wish, countries have the option of storing duplicate seed samples at the Millennium Seed Bank under a Materials Transfer Agreement, which states that the seed remains the property of the donor country.

Huanglong

Given the opportunity, who can resist the chance to see orchids growing in their natural habitats? In 2007, following in the footsteps of the famous plant collector Ernest H. Wilson in the early 1900s, a number of participants from the OSSSU China workshop travelled across the Chengdu plain to the foothills of the mountains of northern Sichuan. On the way, they stayed overnight in the ancient walled city of Songpan, where Wilson had reported seeing thousands of the dark maroon *Cypripedium tibeticum*. The China workshop was held in November. It was cold. The hotels didn't believe in heating. Having breakfast in the hotel in Songpan while wearing coats, hats, and scarves was a new experience. Some months later in 2008, an earthquake would devastate towns along our route, and tens of thousands of people would lose their lives.

It is difficult to imagine a more beautiful valley than Huanglong, with its backdrop of snow-capped peaks of the Min Shan range, and Xuebaoding rising to 5,588 m (18,30 ft). At the head of the valley bubbles a geothermal spring, its waters rich in calcium carbonate minerals that deposit a bed of travertine rock. As the calcium carbonate precipitates, it forms a series of terraces of crystal-clear pools in shades of celestial blue and turquoise. Whereas first-time visitors from temperate countries to tropical lowlands in some parts of the world can struggle with the unaccustomed heat and humidity, on this trip our colleagues from Indonesia and Thailand were wrapped up like Michelin men. Because of the thin air at that altitude, visitors are offered oxygen along the route.

The Hengduan Mountain Region is a global biodiversity hotspot, and the Huanglong Valley, with its rich temperate/alpine flora, is home to more than 100 orchid species. None of them were in flower that late in the year, but seed biologists get excited by seeing seed capsules! Evident beneath the open canopy of the shrubby vegetation, which is typical of thin soils on limestone, were the brown remnants of *Cypripedium flavum*. Wilson had collected hundreds of this species in the region. A report in the 1913 volume of *The Orchid Review* tells us: 'The roots were dug up in October, 1910, transported some eighteen hundred miles by porters, boat, and steamer, and finally shipped from Shanghai on March 14th, 1911. They were received by the Arnold Arboretum, Boston, on April 12th.' Remarkably, some of the plants grew and flowered.

Quito to Papallacta
In June of 1802, Alexander von Humboldt and his companion, Aimé Bonpland, began to climb Chimborazo Volcano in Ecuador, which rises to more than 6,200 m (20,300 ft) and was, at that time, thought to be the world's highest mountain. Taking detailed measurements wherever he went, Humboldt was the first person to undertake a thorough scientific investigation into factors that affect plant distribution. Noting how the flora changes with increased elevation, he divided the vegetation into altitudinal bands or 'life zones', where different plant communities grow according to temperature and humidity.

Today it is possible to travel along the road from Quito to Papallacta and beyond and see for yourself how the vegetation, including the orchid flora, changes with altitude. Participants in the Quito workshop followed the same cobbled road out of Quito that Spanish conquistador Gonzalo Pizarro took with his army in 1541, up into the Andean páramo. Here, at 4,100 m (13,450 ft), a mere stone's throw from the equator, snow often carpets the ground. Amidst the clubmosses and a distinctly Alpine flora was found *Aa hartwegii*, resembling a brown shrivelled piece of straw, more dead than alive.

Making their way down the eastern slopes of the Andes towards today's Papallacta and Baeza, Pizarro and his followers must have seen Andean condors (*Vultur gryphus*). Today, fewer than 100 of these huge soaring birds remain in Ecuador. Pizarro battled his way down mountainsides that were cloaked in thick forest. Only fragments now remain of the high-altitude ancient *Polylepis* forests, home to tiny *Lepanthes* species. Evidence of recent geological activity is everywhere as you drive through the Antisana National Park. Antisana volcano last erupted in 1801/02, and the rough and rocky *pedregales* (old lava fields) are home to an array of orchid species.

Wherever the ground is flat, the original forest has been replaced by pasture with occasional trees spared to provide shade

Francisco Tobar photographing *Telipogon* species on an isolated tree in a pasture on the western slope of the Andes in Ecuador.
Photo: Philip Seaton

for livestock. At the first stop, climbing gingerly over the barbed wire fence, participants were treated to oncidiums and telipogons sheltering in the branches of the trees. Our young guide, Francisco Tobar, obligingly climbed one of the trees to photograph the telipogons. In common with European 'bee orchids', they offer pollinators no food reward. Adam Karremans tells us that the centre of the flower's uncanny resemblance to a female fly entices the male insects to visit. Further along the road, in a pasture where

282 SAVING ORCHIDS

(opposite top)
'Doomed' telipogon: once the forest remnant has been cleared, the orchids will disappear along with it.
Photo: Francisco Tobar

(opposite bottom)
Not all orchids are beautiful. There is little hope of saving a species like *A. hartwegii* (here photographed at 4,100 m (1,345 ft) above sea level) if the more charismatic orchid species can't be conserved.
Photo: Francisco Tobar

the original vegetation clung to the banks of a stream, were tiny lepanthes thriving in the humid air.

A lifelong passion

It is difficult to overstate the importance of Malmgren's medium for the propagation of terrestrial orchids. Today it is used around the world by amateur and professional growers to propagate many species that were once regarded as either difficult or impossible to grow. How Svante Malmgren came to develop his medium is a lesson for scientists and non-scientists alike.

His passion for orchids began when he was a child growing up in the Swedish countryside, in what he describes as a botanical heaven, with orchids growing behind the house and in the surrounding area. On his 15th birthday he was given a copy of *Orchids of Europe* by Aloys Duperrex. The book included a short chapter about the propagation of tropical orchids, with a note that said, 'it seems possible that similar principles may be feasible with some of the European orchids.' Malmgren was hooked.

The following year, he sowed seeds of *Dactylorhiza* and *Gymnadenia* species on a home-made medium. He had an old pressure-cooker, and could only afford to buy two Erlenmeyer flasks. His first attempt led to minimal germination, and the seedlings eventually died because he didn't know at the time that the medium could dry out, or that seedlings needed fresh medium. Undaunted, he repeated the process again and again, getting slightly better results each time. Later, after reading articles by Werner Frosch in *Die Orchidee*, he realised that European orchids really could be propagated from seed. Through *Die Orchidee,* he was able to get in touch with some of the early orchid pioneers in Germany, most notably Gertrud Fast. She had tried different 'organic complexes' and suggested that he incorporate pineapple juice into his media. Svante says that he owes her a continuing debt of gratitude for her support.

As a medical student he learned sterile techniques and, using TGZ medium and pineapple juice, he continued to try out different media combinations. He had a lot of success with many *Dactylorhiza* species, some *Ophrys*, but not *Cypripedium*. At this stage, despite carrying out more experiments using home-made media, there was generally no improvement. The breakthrough came when he decided to try a different nitrogen source. When hospital patients are dehydrated, they are sometimes put on an intravenous drip consisting of a sugar solution and a little salt (sodium chloride). People who are malnourished also need protein – sometimes given as a proprietary solution containing amino acids called 'Vamin', which contains the 18 amino acids necessary for long-term intravenous nutrition. For paediatric use, a slightly different composition is used, called 'Vaminolact'.

Pineapple juice plus 'Vaminolact' opened many doors.

Almost everything germinated and grew successfully, apart from a few cypripediums, cephalantheras, and other species we've previously called 'problem children'. Malmgren discovered that the mineral base could be considerably simplified, until finally no micronutrients were required (except those that were present in the pineapple juice) . . . and just tap water. He now had 1,500 to 2,000 Erlenmeyer flasks, allowing him to experiment, often by trial and error, sometimes with great success. At roughly the same time, he found that most Mediterranean species grew very well on his media. He had excellent growing facilities consisting of a cool cellar and a big, old, frost-free cow shed, where he could talk all winter to what he refers to as his 'little orchid patients'. Eventually the collection grew to a few thousand plants.

Gradually Malmgren came to understand the importance of temperature when growing European orchids. Unlike their tropical cousins, European orchids are subject to large temperature differences throughout the changing seasons, which clearly have a major impact on the growing cycle. The aim is to obtain healthy, mature little tubers at the right time of year, which allows them to be successfully transferred to soil. Svante tells us that this can't be emphasised or repeated too often to beginners and non-beginners alike: 'You can use many different media, but you must sow the seed at the right time of the year. You need to work out if it is a species that also needs a cool period whilst growing on medium in the flasks, or very cool after germination . . . just as in nature. If you are having problems, that is the golden key. Orchid protocorms and small plants have a genetic clock, but this one is also governed by temperature sequence to work!'

Malmgren has, so far, successfully propagated around 200 different orchid species and hybrids, and he has donated plants to friends and orchid growers over many years, mostly to conservation projects. In most cases, reproducible methods have been developed for large-scale production. It must be understood, however, that certain species remain tricky to grow. Poor rates of germination and less than perfect growth on media can still be a problem, but these issues are likely to be solved by ongoing trials. In addition, some orchid genera and species have not yet been tried at all. Some interesting challenges remain for future generations of orchid growers.

What have we learned from Malmgren? From talking with him and other orchid enthusiasts, it is clear that early exposure to the natural world often leads to a lifelong passion for nature. We all owe a debt of thanks to those who came before, and to those in the orchid community who are generous in sharing their experiences. You don't need sophisticated laboratory facilities. Patience and persistence are the keys to success. Malmgren's empirical approach of trial and error, together with a willingness to experiment and to learn from other people's experience, have led to his ability to germinate the seeds of most European terrestrial orchids.

Storing fungi

The use of asymbiotic techniques has generally been so successful that we often forget that mycorrhizal fungi serve this purpose in nature, and are vital for orchid conservation. The reality is that if certain fungi were to disappear in nature, orchids would be unable to spawn seedlings without human intervention. All orchid seeds depend on a dangerous liaison with a fungus to germinate. The bird's-nest orchid remains dependent on its fungus throughout its lifetime, only appearing above ground to flower. Other species may require different fungi at different stages in their life cycles, or perhaps entirely shed the need for any fungi whatsoever at maturity. Storing fungi, like storing orchid seed, is finally, at the forefront of our conservation efforts.

Compared to all other plants that form mycorrhizal associations with fungi, orchids are fortunate because the fungi they utilise are, for the most part, free-living entities that feast on decomposing organic matter. For this reason, orchid fungi are easy to keep alive in storage independently of the orchid. Because most are basidiomycetes – higher fungi akin to mushrooms – most can be stored in a viable condition at $-196°C$ ($-320°F$), and revived when needed. In Canada, the UAMH Centre for Global Microfungal Biodiversity contains many types of cryopreserved orchid mycorrhizal fungi. It serves as the largest public Microbial biobank in the Western Hemisphere, with more than 12,000 accessions of living fungi and bacteria in storage. Orchid mycorrhizal fungi are also conserved in similar conditions at Kew, and there are collections elsewhere (for example in Ecuador, China, and Australia). Ideally the fungus and seed would be stored as one unit, and currently progress is being made to co-culture and store orchid seeds and their mycorrhizal fungi together in alginate beads at refrigerator temperatures, so that both organisms can be quickly revived when removed from storage, with the additional benefit of avoiding the need for frequent subculturing of the fungus.

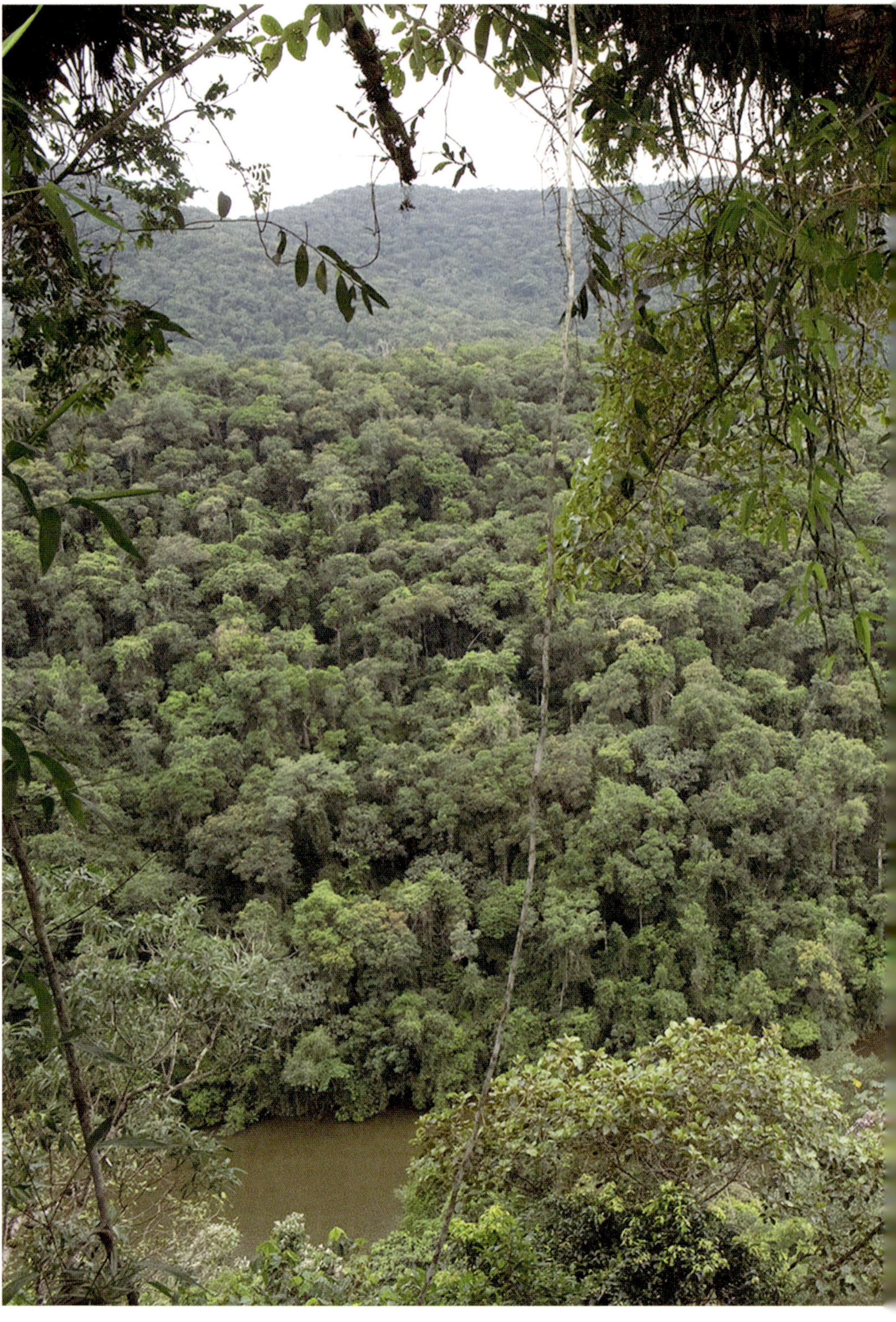

19 The way ahead

'Never doubt that a small group of thoughtful, committed citizens can change the world; indeed, it's the only thing that ever has.'
Margaret Mead

The path ahead may be illuminated by the small nation of Costa Rica. Situated on a narrow isthmus of the Central American 'land bridge' between North and South America, and occupying just 0.3% of the Earth's surface area, Costa Rica harbours almost 6% of the world's biodiversity. Robert Dressler estimated the number of orchid species found there at around 1,200, the majority of them epiphytes, rooted on trees bathed in moisture in hills and mountains below 2,000 m (6,500 ft).

On 23 April, 1925, Aimé Tschiffely set out from Buenos Aires, Argentina, with two horses, named Mancha and Gato, and travelled overland 17,000 km (10,000 mi) to New York, USA. People thought he was crazy. This was the heyday of the Model T Ford, but paved roads throughout South and Central America were few and far between. The future Inter-American Highway was just a dream. Much of the journey was along hazardous mountain trails that led through dense tropical forest. Riding was often impossible, and Tschiffely relates that in Ecuador there were, 'places where one had to consider oneself lucky if the animals can pass without accident.' On reaching the Pacific coast of Costa Rica, he was faced with crossing the infamous 3,341 m (10,961 ft) Cerro de La Muerte – the 'mountain of death' – to San José in the Central Valley, a gruelling journey that involved stopping for two nights at government shelters. Why was it called the 'mountain of death'? Travellers journeying from the tropical heat of the lowlands who were not adequately prepared would often succumb to cold and rain at the mountain's peak. Nowadays, it is possible to travel by road from Costa Rica's Pacific to its Caribbean coast in a matter of hours.

In the 1940s, around three quarters of the country was still covered by forest, but by the 1980s, cover was reduced to around 21%. This was in part due to the conversion of forest to pasture for raising cattle to produce cheap beef for the USA – the 'hamburger connection'. Then Costa Rica decided to follow a different path and protect its natural heritage, increase levels of bio-literacy and educate its children accordingly. Today, 60% of Costa Rica's land is cloaked by forest spanning all 12 ecological zones. Nearly one-third of its territory has been set aside as national parks, providing the country with a flourishing ecotourism industry. Costa Rica is special because the people and its elected leaders cherish its biodiversity. Recognising the importance of the participation of local citizens, who are most familiar with their local ecosystems, has been one of the keys to the country's success.

(previous pages)
Using modern secure climbing techniques, today's orchid biologists are able to climb trees in pursuit of more understanding of the ecological requirements of orchids and other accompanying epiphytes.
Photo: Luciano Candisani

Sobralia growing in a tree in the Monteverde Cloud Forest, Costa Rica.
Photo: Philip Seaton

A global recipe for success?

During the past 3.8 billion years there have been at least five mass extinctions, and few question that we are currently witnessing Earth's sixth mass extinction. It has been estimated that somewhere between 30 and 50 species become extinct every day, and this number is expected to accelerate as the world continues to warm. Political leaders participating in the United Nations Summit on Biodiversity in September 2020, representing 78 countries from all regions, committed to reversing biodiversity loss by 2030. Despite the warm words, a sense of urgency is still lacking, and in most countries, action to conserve biodiversity lags behind action to address the climate crisis.

But how should we as individuals respond to the biodiversity crisis? What kind of world do we want to leave our grandchildren and the generations that follow? Sadly, we will never be able to reclaim the abundance of the past. The IUCN Orchid Specialist Group is a global network of experts who volunteer their time and expertise with the aim of building a scientific and practical foundation for orchid conservation. Indeed, many of the people who have contributed to this book are members of this august body. However, conservation is not, and should not, be seen as the province of experts alone. For example, there is a vast untapped reservoir of knowledge and expertise residing in the amateur orchid-growing community. The concept of 'citizen scientists', where non-scientists are encouraged to participate in various projects supported by working scientists, is rapidly gaining popularity. It would be wonderful to think that we could form a group of 'citizen orchid conservationists', perhaps with National Orchid Collections at their core. While there is no magic wand that we can wave to resolve everything, there are solutions to individual problems. The key often lies with a local community conserving an individual species, such as *Cattleya mossiae* in Venezuela, or with a committed group of people, as exemplified by the Wildlife Trusts in the UK, who are conserving habitats and the orchids they contain.

The internet has transformed the ways in which information is accessed, providing new opportunities for reaching and involving a wider audience, including the many young people who are searching for ways to participate in orchid conservation. Orchid societies can play a pivotal role in sharing information – through journals, talks, workshops, and at shows – not only about cultivation, but also about conservation issues, including poaching. Part of the answer lies in educating the general public about the negative effect that the collection of plants is having on wild populations. Growers can learn more about the species they are cultivating in their collections. Is it a species at risk? With care and attention, many orchid species can live for many decades. Who knows what gems may be lurking in orchid collections? Growers have a special responsibility if their collection houses rare and endangered plants. The plants' futures in living collections can be ensured through propagation and sharing them with other growers.

People can be encouraged to make provision for their plants when they are no longer able to care for them. Michael McIllmurray's legacy lives on at Kew. His National Collection of *Maxillaria* species is now part of Kew's living collection, seeds resulting from pollinations he carried out in his greenhouses are safely stored at the Millennium Seed Bank, and his herbarium specimens and associated collections now reside in Kew's herbarium as an important resource for future investigators.

The example of Beth Otway teaches us that you don't need a large collection to make a worthwhile contribution. With limited space, she grows her plants in a terrarium where she houses the National Collections of miniature *Phalaenopsis*, *Aerangis*, and *Angraecum* species. After hand-pollinating her plants, she sends the resulting seed to enhance collections in botanical gardens such as Kew.

Orchid conservationists frequently invest considerable time and money travelling to conferences in faraway places, where they mingle and share ideas on how to preserve what remains of our planet's dwindling biological heritage. After the meeting, everyone leaves for home energised and firmly convinced that there must be a 'call to action'. Invaluable as these meetings are, there remains a problem: the participants have been largely 'preaching to the choir'. How can conservationists reach and engage with a wider audience? Why not, for example, put on educational displays at the Chelsea Flower Show? Held in London each May, Chelsea is arguably the world's most prestigious flower show, drawing exhibitors from around the world. In 2023, John Parke Wright IV of Florida came up with the idea of promoting global orchid conservation through 'Orchid Conservation Chelsea', an annual series of displays highlighting orchids from different countries around the world. Thus a team of orchid specialists was assembled from the USA and Kew to create the first exhibit, showcasing the native orchids of Britain, the Republic of Cameroon, and the US State of Florida. The main attraction was to be Florida's ghost orchid. The display's success hinged on finding and transporting a ghost orchid that had originated in cultivation, and would be in bloom in time for Chelsea, two months before it typically flowered in nature. Enter Johanna Hutchins, the Chicago Botanic Garden's orchid floriculturist. As time ticked by, one of the ghosts in her care began to show signs of a bud. A CITES permit was obtained well ahead of time. Phytosanitary inspections were successfully negotiated and, carefully cradling the plant in her lap throughout the journey across the Atlantic, Hutchins delivered the precious plant in time for it to take centre stage in the exhibit.

    The backdrop to the display was a stunning black-and-white image of the Everglades, taken by legendary photographer Clyde Butcher. A path led visitors through the exhibit, allowing them to enjoy the beauty and diversity of orchids while learning more about the importance of conserving their natural habitats. Although the interest of many focused on the ghost orchid, others were equally interested in the British native orchids and the orchids of Cameroon, which were probably seen for the first time at Chelsea. A unique feature was the 'genius bar', where orchid experts were able to field questions. Visitors were able to examine a 3-D printed ghost orchid up close, and to take away a printed 'Orchid-Gami' ghost orchid produced by the North American Orchid Conservation Center to make at home. A small case contained the ghost's long-tongued moth pollinators, in addition to flasks of seedlings raised by students at Fairchild Tropical Botanical Garden in Miami.

    At the front of the stand, a large specimen tree was draped with Spanish moss (*Tillandsia usneoides*), and other bromeliads were planted in its branches to resemble the Florida Everglades. Twelve leafless orchid ghosts of various sizes were placed at strategic locations in the tree, and visitors were invited to find them, highlighting the difficulty that researchers have in spotting them in their natural habitats. Hutchins' larger, budding ghost was planted in the centre of the exhibit, making it easier to find – at least it was when it was pointed out by one of the team!

Never has an orchid been watched with so much anticipation. Initially the weather was decidedly cool, but as the days went by the temperature in the marquee increased. Would the orchid oblige, and the flower bud open at Chelsea? The team placed a hot water bottle beneath the ghost during one cool night. The next day, the hot water bottle was replaced by chemical heat packs wrapped in Spanish moss. The orchids were regularly sprayed with water to prevent the roots drying out, but in the end, the atmosphere in the marquee was too cool and dry for the flower to open.

All was not lost. After the show, the plant was donated to Kew and displayed in a terrarium in the Princess of Wales Conservatory. Two days later, the bud opened – and created a media sensation. Visitors came in their droves to see the now famous ghost orchid flowering in the UK. The story appeared in newspapers, social media, and on TV in the UK and around the world, effectively calling attention to orchid conservation on a global stage.

The stories told in this book are a celebration of the positive actions being taken by numerous groups around the world. We began with the story of Florida's ghost orchid: to show how a small team of dedicated individuals, in this case students and researchers, joined forces to save a charismatic species from possible extinction in our rapidly changing world, overcoming hurdles ranging from reptiles to politicians. The story demonstrates how an integrated approach, involving both *in situ* conservation (conservation of orchids within their natural habitats) and *ex situ* conservation (the use of laboratory techniques), can be equally important in achieving success. We have been humbled by the willingness of friends and colleagues to share their own stories, to write of their personal experiences – in truth there have been too many to include in one book. We have been awestruck by the many success stories throughout the world – and by the enthusiasm of an increasing number of young people who are joining the fight to save orchids and their habitats, many of whom have already made a positive impact. Our wish is that future generations will be able to look back at this time in history, in a world filled with the same orchids and other life forms that we grew up with – and smile.

Silver medal winning exhibit of orchids of Florida, Cameroon and Great Britain, featuring a Florida ghost orchid growing in tree strung with Spanish moss (*Tillandsia usneoides*) at the 2023 Chelsea Flower Show.
Photo: Joyce Seaton

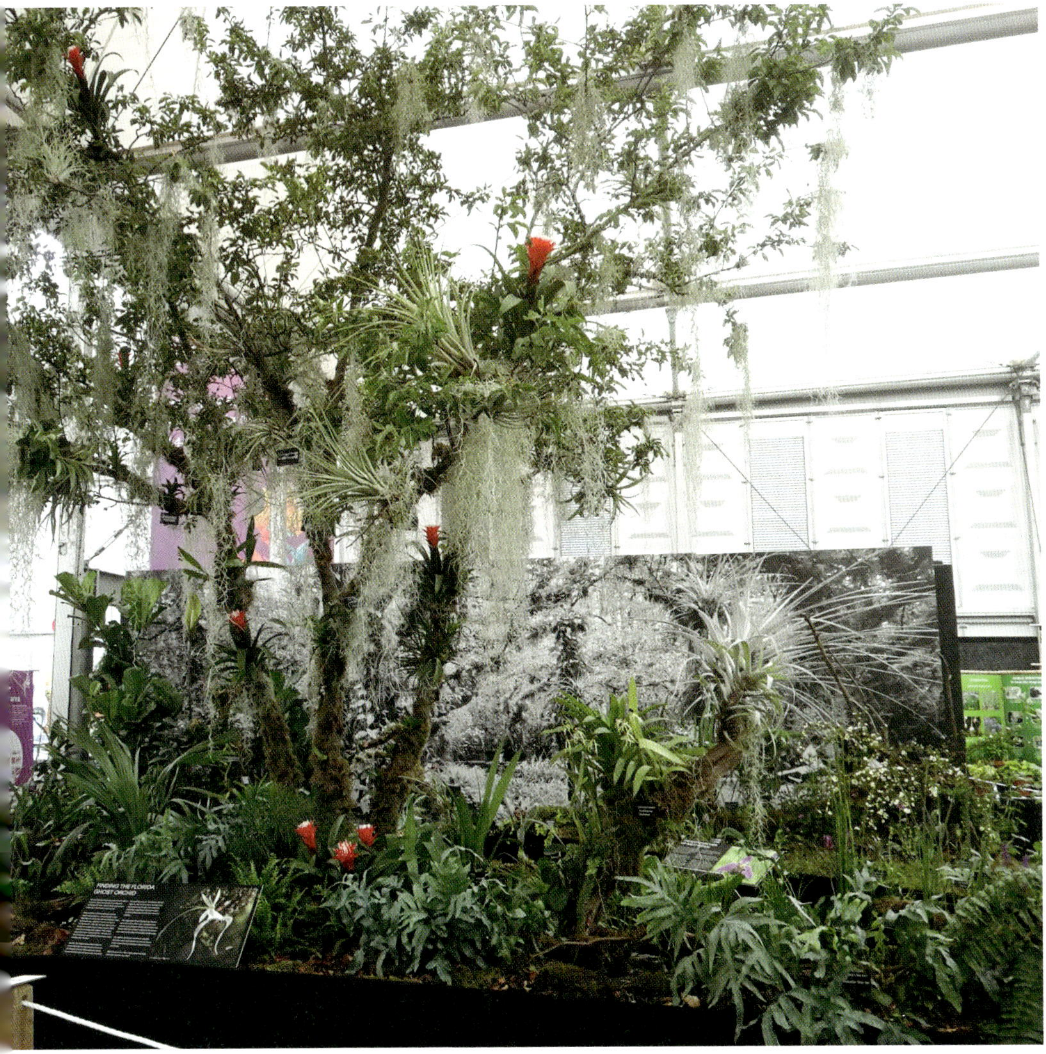
FINDING THE FLORIDA
GHOST ORCHID

Further reading

Antonelli, A. et al (2023). *State of the World's Plants and Fungi 2023: Tackling the Nature Emergency: Evidence, Gaps and Priorities*. Royal Botanic Gardens, Kew.
Bersweden, L. (2018). *The Orchid Hunter: A Young Botanist's Search for Happiness*. Short Books Ltd., London.
Danaher, M. W., Ward, C., Zettler, L. W. & Covell, C. V. (2019). Pollinia removal and suspected pollination of the endangered Ghost Orchid (*Dendrophylax lindenii*) by various hawk moths (Lepidoptera: Sphingidae): another mystery dispelled. *Florida Entomologist* 102 (4): 671–683.
Dixon, K. W., Kell, S. P., Barrett, R. L. & Cribb, P. J. (eds) (2003) *Orchid Conservation*. Natural History Publications, Borneo.
Dressler, R. L. (1981). *The Orchids: Natural History and Classification*. Harvard University Press.
Dunn, J. (2018). *Orchid Summer: In Search of the Wildest Flowers of the British Isles*. Bloomsbury Publishing, London.
Karremans, A. P. (2023). *Demystifying Orchid Pollination: Stories of Sex, Lies and Obsession*. Royal Botanic Gardens, Kew.
Koopowitz, H. (2001). *Orchids and their Conservation*. B.T. Batsford Ltd., London.
Lee, Y.-I. & Chee-Tak Yeung, E. (eds) (2018). *Orchid Propagation: From Laboratories to Greenhouses – Methods and Protocols*. Humana Press, New York.
Luer, C. A. (1972). *The Native Orchids of Florida*. New York Botanical Garden, New York.
Manning, S. (2010). *Discovering New World Orchids*. Published by Steve Manning.
Miller, D., Warren, R., Miller, I. M., & Seehawer, H. (2008). *The Organ Mountain Range, Its History and Its Orchids*. Editora Scart, Nova Friburgo/RJ, – Brazil.
Millican, A. (1891). *Travels and Adventures of an Orchid Hunter*. Literary Licensing LLC, Montana, USA.
Orlean, S. (1998). *The Orchid Thief: A True Story of Beauty and Obsession*. Random House, New York.
Rasmussen, H. N. (1995). *Terrestrial Orchids: From Seed to Mycotrophic Plant*. Cambridge University Press, Cambridge.
Seaton, P. & Ramsay, M. (2005). *Growing Orchids from Seed*. Royal Botanic Gardens, Kew.
Seaton, P. T. (2023). *Odontoglossum crispum*: a tale of love, loss and scientific discovery. *Lankesteriana* 23 (3): 593-612.
Seaton, P., Cribb, P., Ramsay, M. & Haggar, J. (2011). *Growing Hardy Orchids*. Royal Botanic Gardens, Kew.
Swarts, N. D. & Dixon, K. W. (2009). Terrestrial orchid conservation in the Age of Extinction. *Annals of Botany* 104: 543–556.
Swarts, N. D. & Dixon, K. W. (2017). *Methods for Terrestrial Orchid Conservation*. J. Ross Publishing, Florida.
Swinson, A. (1970). *Frederick Sander: The Orchid King*. Hodder and Stoughton Ltd., London.
Zettler, L. W. & Perlman, S. P. (2012). *Peristylus holochila* – Hawaii's rarest orchid and the battle to save it from extinction. *Orchids* 81 (2): 94–99.
Zettler, L. W. & Seaton, P. (2023). A ghost story. *Orchid Society of Great Britain Journal* 72 (3): 226–233.
Zettler, L. W., Kane, M. E., Mujica, E. B., Corey, L.L. & Richardson, L. W. (2019). The ghost orchid demystified: biology, ecology, and conservation of *Dendrophylax lindenii* in Florida and Cuba. In: Pridgeon, A. M., Arosemena, A. R. (eds), *Proceedings of the 22nd World Orchid Conference*, September 2019. Guayaquil, Ecuador Vol. 2: 136-148.

Acknowlegements

Writing this book has been a fascinating journey. Many wonderful people have made it possible by sharing their expertise, knowledge, and above all, their stories. Any mistakes or inaccuracies, however, are entirely our responsibility.

We are aware that there are many more people who have made this book possible, and whose names appear in the text. What follows are the names of just some of those people, and we apologise to those whose names do not appear. Jennifer Zettler, Bob Fuchs, Promila Pathak, Maria Elena Cazar Ramirez, Raffaella Ansolani, Lydia Ballard, Vicente Perdomo, Leticia Abdala, Diego Bogarín, Sandro Cusi, Eduardo Pérez, Catalina Restrepo, Yu Zhang, Tom Mirenda, Camilo Uribe Botta, Johan Hermans, Danielle Ferreira, Matt Richards, Stig Dalström, Richard Warren, Ingrid Morales, Ernesto Mujica, Tim Wing Yam, Rebecca Hsu, William Cetzal Ix, Yung-I Lee, Dwi Murti Puspitaningtyas, Kanchit Thammasiri, Samuel Sprunger, Ken Cameron, Svante Malmgren, Edgar Mo, Michael Fay, Lou Jost, Sebastian Vieira, Marta Kolanowska, Wenqing Perner, Marc Freestone, Noushka Reiter, Sergio Elórtegui Francioli, Rodolfo Solano, Jason Downing, Roger Hammer, Ana María Sánchez-Cuervo, Justin Kondrat, Michele Lussu, Alexandre Antonelli, Jorge Warner, Tim Marks, Louise Colville, Udayangani Liu, Gerardo Castiglione, Clyde Butcher, Benjamin Crain, Carol and F. William Zettler, Mark Danaher, Larry Richardson, Kit and La Raw Maran, Brent Chandler, Jeff Norris, Adam Herdman, Pilar Ortega-Larrocea, Mike Kane, Viswambharan Sarasan, Jonathan Kendon, Kazutomo Yokoya, Hank Oppenheimer, Steve Perlman, Andrew Stice, Landy Rajaovelona, Elaine González, José Bocourt, Haleigh Ray, Hoang Nguyen, Hanne Rasmussen, Rachel Helmick, Andrés Ernesto Ramos Roldán, John Shaw, Samantha Koehler, Antonio Ruiz, Ana Maria Benavides, Andrés Ramos, Nelson Barbosa Machado-Neto, Svante Malmgren, Luciano Zandoná, Hugh Pritchard, Lou Jost, Emily Coffee, Elizabeth Rellinger Zettler and Audrey Zettler.

Special thanks to Claire Seaton, Phillip Cribb, Julianne McGuinness, and Dennis Whigham, who read through the earlier versions with such care and made some valuable suggestions. We are especially indebted to Georgie Hills and Lydia White, without whose help and guidance this book would not have been possible, and James Kingsland, whose edits and thoughtful comments have been invaluable.

Dedication: Philip Seaton
For Ted, Finch, and Cary.

Dedication: Lawrence Zettler
To my parents, Bill and Carol, who cultivated my interest and concern for the natural world every step of the way.

Index

Photos are shown in **bold**

A

Aa hartwegii, 280, **282**
Abdala, Leticia, **254-5**, 260
Acineta barkeri (bumblebee orchid), 164, **166**
Ackerman, Jim, 36, 37
acid(s),
    hydrochloric, 101
    lactic, 31
    malic, 31
    organic, 37
    picric, 58
    rain, 143
    threonic, 31
Adaptation (movie), 17
Aerangis, 290
    A. ellisii, 81, 82
Aerides, 114
Africa, 34, 38, 49, 62, 75, 116, 159, 196
agar (see media)
AK-47, 79
Alaska, 68, 69
    Aleutian Islands, 68
alcohol, 84, 103
    4-hydroxyl benzyl, 31
Alexander the Great, 62
algae, 22
alginate beads, 285
alkaloid, 245
alligator(s) (*Alligator mississippiensis*), 14, 19
Alpine Garden Society, 237
Alto de Ventanas, 208, 210
Amazon, 135, 155, 170, 215
amber, 30
American Orchid Society, 271
amino acids, 283
Anacamptis,
    green-winged orchid (*A. morio*), 140, **179**, 244
    lax-flowered orchid (*A. laxiflora*), 237
    pyramidal orchid (*A. pyramidalis*), 140, 246, 249
André, Eduard, 105
Andriantiana, Jacky, 80, 81
Angraecum, 290
    A. cadetii, 32
    A. longicalcar, **76**, **77**, **78**, 79, 82, **132**, 260
    A. sesquipedale, **27**, 32
    A. sororium, **133**
anguloas, 213
Animal and Plant Health Agency, 151
Antarctica, 62, 155, 180
Antarctic plants,
    Colobanthus quitensis (a pearlwort), 180
    Deschampsia antarctica (a grass), 180
anther cap, 30, **211**, **272**
anthuriums, 192, 250
    Anthurium caramantae, 250
Antonelli, Alexandre, 177
ants, 57, 83
    Azteca, 200
    Camponotus, 200
aphids, 32, 264
aphrodisiac, 112, 116, 119
Apostasia nipponica, 34

arandas, 113
Arctic tundra, 30
Argentina, 171, 288
    Buenos Aires, 288
Army Corps of Engineers, 15
Arnold Arboretum, 280
aroids, 207
artificial intelligence (AI), 185
Arundina, 35
ascocendas, 113
Asia, 34, 35, 36, 62, 73, 74, 75, 134, 155, 167, 196, 198, 230, 238, 276, 278
    Southeast, 36, 63, 73, 192
Asteraceae, 176
Atlanta Botanical Garden, 39, 100, 223, 225, 274, **275**, 276
Atlantic Forest, 170, 171, 172, 175, 177, 222
Atlantic Forest Research and Conservation Alliance (ARAÇÁ), 177
Atlantic Ocean, 108, 151, 155
Australia, 32, 34, 50, 62, 63, 66, 74, 83, 84, 85, 86, 87, 90, 91, 95, 124, 135, 152, 158, 210, 235, 237, 244, 277, 285
    bushland, 84, 85
    Kings Park and Botanic Gardens, 83, 146
    Native Orchid Society (Victorian Group), 87
    New South Wales, 95, 124, 155
    North Queensland, 36, 83
    Perth, 83, 84, 85, 235, 244
    Royal Botanic Gardens, Victoria, 86, 87
    Tasmania, 180
    University of Melbourne, 91
    Western, 83, 84, 85, 146, 188
Australian Network for Plant Conservation, 155
autumn coral root (*Corallorhiza odontorhiza*), 243
Averyanov, Leonid, 230

B

bacteria, 22, 39, 274, 275, 285
bananas, 30, 50, 119, 225
banded flower mantis (*Theopropus elegans*), **251**
bandicoots, 83
bandits, 80
Bangladesh, 155
Baltic, 30
bark (trunks), **14**, 19, 21, 22, 23, 24, 37, 38, 39, 177, 207, 222, 223, 252
Barrett, Russell, 235
basidiomycetes (see mushrooms)
Bateman, James, 51, 52, 56, 58, 104, 105
bats, 176, 209
bears, 217
beef (see cattle)
bee(s), 30, 135, 153, 166, 172, 175, 176, 192, 200, 208, 249
    bumblebee, 164, 174
    carpenter, 176
    Colletes cyanescens, **33**
    euglossine, 42, 176, 200, 201, **203**
    megachile, 253
beetles, 28, 208, 209, 257
Belgium,

Bruges, 57
Pauwels, 59
Bernard, Noël, 181
Besi, Edward Entali, 132
Bhutan, 126, 129, 278
Bipinnula fimbriata, 192, 193
    pollination, **33**
    rescue, **195**
bilby, 83
bilharzia, 82
biodiversity hotspot(s), 15, 84, 94, 129, 130, 131, 171, 180, 183, 190, 192, 206, 209, 210, 215, 280
biological control, 159
biosensor, 99
binomial nomenclature, 98, 99
birds, 32, 34, 39, 49, 63, 66, 67, 69, 70, 74, 138, 143, 175, 176, 188, 206, 208, 209, 217, 218, 230, 253, 280, 285
    Andean cock-of-the-rock (*Rupicola peruviana*), 217
    Andean condor (*Vultur gryphus*), 280
    bee hummingbird, zunzuncito (*Mellisuga helenae*), 106
    black-and-chestnut eagle (*Spizaetus isidori*), 217
    Cuban trogon, tocororo (*Priotelus temnurus*), 106
    plate-billed mountain-toucan (*Andigena laminirostris*), 217
Black, J. M., 182
black market, 146, 149
Blechnum, fern, 158
Bletia urbana, 143
Bogarín, Diego, 102, 129, 158
Bone, Ruth, 118
bonsai, 113
Bolivia, 95, 276, 278
Bonpland, Aimé, 49, 280
Borneo, 70, 210
Botanic Gardens Conservation International, 256
Boyle, Frederick, 38, 49, 50, 51, 53, 57
Brazil, 10, 11, 62, 105, 170, 171, 172, 175, 177, 199, 201, 206, 207, 222, 225, 270, 276
    Campinas, University of, 177, 223
    Cantareira, 146, 171, 172, 173, 222, 223
    Macaé de Cima, 171, 175, 176
    nuts, (*Bertholletia excelsa*), 42
    Pernambuco, 108
    Presidente Prudente, 199
    Rio de Janeiro, 107, 108, 135, 175, 209
        Botanic Garden, 175
    São Paulo, 146, 147, 172, 175, 177, 222
    Serra dos Orgaños Mountains, 10, **64–65**
    University of Western São Paulo (UNOESTE), 199
British Antarctic Survey, 180
British Museum of Natural History, 181
British orchids, 42, 153, 162, 196, 237, 291
bromeliads, 15, 63, **166**, 172, 176, 192, 200, 207, 208, 213, 217, 225, 252, 291
Broughtonia lindenii, 138
bryophytes (see mosses)
Bulbophyllum, 30, 114

B. kwangtungense, 131
B. medusae, **221**
B. membranaceum, 251
B. vaginatum, 251, 252
Bull, William, 50
bumblebee orchid (see *Acineta barkeri*)
Burberry, 56
burlap, 23, 24
Burma (see Myanmar)
Butcher, Clyde, 9, **12–13**, 14–15, 291
butterflies, 30, 32, 67, 70, 206, 210, 272
  in Hawaii,
    Blackburn's bluet
      (*Udara blackburni*), 67
    Kamehameha butterfly
      (*Vanessa tameamea*), 67
  in Indonesia, *Ornithoptera*, 70

C
cacao (see chocolate)
cacti, 39, 106, 143, 207, 260
caiman, 49
Caladenia species,
  C. amoena, 90
  C. arenicola, 86
  C. audasii, 87
  C. cretacea, **88–89**
  C. cruciformis, **88–89**
  C. longicauda, 84, **85**
  C. leucochila, 147
  C. procera, 147
calcium chloride, 274
California, 152, 159, 190
Caligari (Sardinia), 140
Calochilus richiae, 90
CAM metabolism, 39
capsule(s), **14**, 35, 72, 79, 108, 117, 120, 142, 150, 151, 166, 182, 199, 225, 237, 238, 239, 249, 250, 252, **261**, 268, 273, 274, 275, 277, 280
  infestation with insects, 72, 159
carbohydrate(s), 37, 91
  starch, 91
carbon, 42
carbon dioxide, 39, 208, 250
Caribbean, 35, 106, 138, 140, 180, 189, 216, 257, 288
carnivorous,
  caterpillar, 66
  plant, 63
carousel spider orchid
  (*Caladenia arenicola*), 86
Cartwright, Ian, 200
cassava, 82
Castiglione, Gerardo, 189
Castilho Custódio, Ceci, 200, **201**
Castle, 58
caterpillar(s), 66, 67
cattle (livestock), 80, 94, 126, 135, 171, 185, 208, 259, 288
Cattley, William, 105, 107
Cattleya, 56, 63, 107, 136, 196, 198, 199, 200, 249, **256**, 260, 264, 270
  C. amethystoglossa, pollination, **272**
  C. aurea, 50, 53
  C. brevicaulis, 199, **201**
  C. caucaensis, 105
  C. citrina, 99
  C. coccinea (syn. *Sophronitis coccinea*), **170**, 171
  C. dowiana, 52, 198, 216, **219**

C. gaskelliana, 189
C. harrisoniae, 136
C. labiata, 105, 107, **108–9**, 199
C. lueddemanniana, 189
C. maxima, 30
C. mendelii, 50, 52
C. mossiae, 51, 189, 256, 290
C. purpurata, **174**, 175
C. quadricolor (syn. *C. chocoensis*), 103, 104, 105, 137, 259, **260–261**
C. schroederae, collection in Colombia, 48, **54**
C. tigrina, 199
C. trianae, **6–7**, 55, 63, 105, 135, 196, 259
C. walkeriana, 199
C. warscewiczii, 50, 52, 198
Cayman Islands, 138, 139
Central America, 63, 144, 158, 192, 230, 257, 288
Ceratobasidium, **22**, 23, 143
Chamberlain, Joseph, 56, 58, 162
Cambronero, Jorge, 129
Cameron, Ken, 181
Catasetum, 28, 42, 200
Ceiba Foundation, 213, 215
centipedes, 57
Cephalanthera longifolia (sword-leaved helleborine), 126, 145, **178**, 182
Cerro de La Muerte, Costa Rica, 138, **156**
Chadwick, Arthur, 260
chainsaw, 167
chalk escarpment, 140
charcoal, 107, 135
Charlesworth, Joseph, 181, 260
Chelsea Flower Show, 151, 291, **292–3**
chikanda, 119, 120, **121**
childbirth, 116
children, 8, 79, 114, 116, 136, 142, 182, **186**, **187**, 188, 190, 192, 193, 194, 197, 209, 250, 288, 289
Chile, 188, 190, 192, 193, 194, 276
  Valparaíso, 192, 193
China, 10, 32, 36, 94, 113, 116, 130, 131, 145, 156, 217, 230, 240, 241, 276, 277, 279, 285
  Beijing, 230, 241, 257
  Beijing Botanical Garden, 240, 241, 243
  Beijing Forestry University, 257
  Chengdu, 276, 279
  Guangxi, 130, 131
  Huanglong, **92–93**, 94, 237, 279
  Huanglong Nature Reserve, 243
  Huanglong Valley, 280
  Kunming Institute of Botany, 277
  Shanghai, 280
  Shenzhen, 233
  Sichuan, 131, 158, 279
  Songpan, 279
  Yunnan, 156, 158, 233
    Forestry Institute, 95
    Southeast Yunnan, 95
  Yachang Orchid Nature Reserve, 116, 130
  Zhongdian, 158
Chloraea,
  C. disoides, **26**
  C. magellanica, herbarium specimen, 102, **104**
chlorophyll, 39, 98, 116, 243
chocolate (cacao), 106, 117, 159
chromatography, 30

chromosomes, 101, 102, 245
  of *Paphiopedilum philippinense*, 101, **102**
Churchill, Winston, 180
chikanda, 119, 120
CITES, 81, 147, 149, 150, 151, 230, 291
  Appendix I and Appendix II 150, 151
citizen scientist(s), 86, 196, 290
citrus, 42
Claes, Florent, 53
Clements, Mark, 86, 237
climate change, 129, 131, 134, 136, 140, 145, 152, 154, 155, 171, 206, 208, 222, 234, 238, 289
climate crisis (see climate change)
clones, 79, 119, 163, 164, 256, 257, 264
clothing, 216
cloud forest, 39, 49, 55, 63, 94, 140, 145, 152, 153, 162, 166, 170, **191**, 192, 207, 208, 209, 211, 215, 273
  Colombian Andes, **204–205**
  Ecuador, **217**
  Monteverde, Costa Rica, **289**
  Taiwan, **160–161**, 167, **168**
coconut, 30, 42
Coelogyne,
  C. cristata, 113
  C. fimbriata, 131
coffee, 42, 50, 103, 107, 108, 119, 129, 135, 162, 164, 171, 208, 209, 222
  Coffea arabica 119
  coffee filter paper, 100, 276
  leaf rust, 119
colchicine, 245
Colman, Sir Jeremiah, 58
Colombia, 8, 9, 129, 130, 135, 136, 196, 206, 207, 209, 214, 215, 217, 218, 234, 256, 259, 276, 279
  Andes, 48, 105
  Antioquia, 207, 210, 217, 249
  Bogotá, 52, 119
  Cali, 260
  Cauca River, 105, 259
  Cauca Valley, 105, 136, 198
  Chocó, 105
  Guadalupe, 8, 196, 197
    school, 196, **197**
  Jardín Botánico Juan María Céspedes, 259, 261
  Magdalena River, 49, 55, 105, 196
  Magdalena Valley, 105
  Medellín, 50, 119, 183, 189, 197, 248, 249
  Neiva, 135, 196
  Pacho, 50, 53, 55
  Parque Arví, 199, 250
  Quebrada El Oro, 210, 211
  Río Dagua, 52
  Río Suaza, 196
  Tuluá, 260
  Tunja, 209
Colombian Orchid Society, 207, 208, 211, 213, 217
column, 30, 200, 230
compost (see humus)
coniferous cloud forest, 167
Concord, New Hampshire, 153
Conservation International, 94, 256
Convention on Biological Diversity (CBD), 151
Cootes, Jim, 73, 74
Corkhill, Peter, **239**

Coronation Meadows Project, 245
Costa Rica, 94, 112, 124, 129, 134, 138, 152, 156, 216, 218, 276, 277, 288, 289
  Cerro de La Muerte, 138, **156**, 158, 288
  El General Valley, 129
  Monteverde Cloud Forest, 152, 153, **289**
  Parque Nacional Corcovado, 129
  San José, 112
  Tapantí National Park, 94
  University of Costa Rica, 276
Coryanthes,
  C. kaiseriana, 200, **203**
  C. vasquezii, 42
COVID pandemic, 210
cowslip (Primula veris), 244
Crain, Benjamin, 75
cranefly orchid (*Tipularia discolor*), 243
Cretaceous, 30
Cribb, Phillip, 75, 234–5, 237
crickets, 34
Crous, Hildegard, 120
cryopreservation, 200
Cryptochilus sanguineus, 32
Cuba, 17, 19, 22, 36, 98, 106, 119, 138, 139, 155, 159, 227, 276, 277
  Ciénaga de Zapata, 106, 227
  Guanahacabibes National Park, 18, 19, 107, 125, 138
  Havana, 19
  Jardin Botanico Orquideario de Soroa, 139, 277
  Pinar del Rio, 106
  Sierra de los Órganos, **64–65**
  Viñales, 138
Cuban trogon, 106
Cuitlauzina pendula, **27**
Cusi, Sandro, 164
Custódio, Ceci Castilho, 200, **201**
cuticle, 39
cyanobacteria, 39
cyclones (hurricanes, tropical storms), 21, 35, 62, 80, 81, 82, 138, 139, 140, 167
  Hurricane Gustav, 139
  Hurricane Iniki, 68
  Hurricane Irma, 226
  Hurricane Ivan, 139
Cycnoches chlorochilon ('el pelicano'), 30
Cymbidium, 35, 84
  C. bicolor subsp. *pubescens*, 251
  C. cyperifolium, 131
  C. ensifolium, 113
  C. finlaysonianum, **220**, 251, 252, 253
  C. serratum, 32
Cypripedium, 42, 94, 144, 156, 158, 182, 230, 237, 240, 241, 243, 283
  C. acaule, 153
  C. calceolus, 145, 235, 238, **239**, 241, **242**, 243
    in Hungary, 239
    reintroduction in Switzerland, 145, **240–241, 242**
    var. *citrina*, 145
  C. dickinsonianum, 230
  C. flavum, **26**, 243, 280
  C. henryi, 243
  C. irapeanum, 114, 144, **157**
  C. macranthos, **228–229**, 240, 241, 243, 257
  C. shanxiense, 241, 243
  C. sichuanense, 243
  C. subtropicum, 32, 95

C. tibeticum, 131, **236**, 243, 279
Cypripedium Committee, 235, 238
Cyrtopodium punctatum (cowhorn orchid), 223, 225, 227
  in Cuba, **224**
  in the Fakahtchee Strand, 223, 226
  pollination, **227**

D

Dactylorhiza, 99, 101, 245, 283
  D. fuchsii (common spotted orchid), 140, 245, **247**
    pollen storage, 257
  D. incarnata, 185
  D. praetermissa, 153
  D. purpurella, 245
  D. sambucina, 185
dahlia, 143
Daniel Boone National Forest, Kentucky, 272
Darling Wildlife Show, 190
Darwin, Charles, 9, 31, 32, 34, 62, 67, 102
Darwin Initiative, 119, 276
Darwinian selection, 199
Dawson, T., 163
Day, John, 52, 58
De Pannemaeker, Pieter, 105
decomposition, 42, 82, 285
Dendrobium, 63, 72, 84, 113, 116
  cut flowers, Thailand, **110–111**
  D. antennatum, 36
  D. bigibbum, **184**
  D. cuthbertsonii, 72
  D. masarangense, 72
  D. officinale, 116
  D. sulphureum, 72
  D. vexillarius, 72
  D. moniliforme, 116, 170
  D. officinale, 116
Dendrophylax, 98
  D. fawcettii, 98, 138, **139**, 180
  D. lindenii, **14**, 15, 18, 19, **20**, 23, 125
    collecting fragrance, **31**
    collecting nectar, **32**
    in Fakahatchee Strand, **14**
    on burlap, **25**
    pollination, **20–21**
    seed capsule, **14**
    spur length, **106**
  D. porrectus (syn. *Harrisella porrecta*), 42
Department for Environment, Food, & Rural Affairs, Defra, 151
Dichromanthus,
  D. aurantiacus, **142**, 143
  D. coccinea, 143
Didymoplexis stella-silvae, 98
digital camera traps, 19, 28, 149, 217
digital copies, 103
dinosaurs, 30
  Age of, 30, 171
  sauropods, 171
Disa, 120
  D. uniflora (Pride of Table Mountain), 32, **33**, 190
disease(s), 119, 151, 256
  Fusarium wilt, 119
Diuris, 90
  D. callitrophila, 95
  D. fragrantissima (sunshine diuris), 90, 91
  D. magnifica (donkey orchid), 86
  D. punctata, 91

Dixon, Kingsley, 15, 83, 84, 180, 235, 244, 277
DNA, 98, 100, 101, 120
  barcoding, 100, 120
  fingerprinting, 238
  sequencing, 30, 91, 98, 101
Dodson and Gentry, 134
Dominican Republic, 30, 276
donkey orchids (see *Diuris*)
Dorothy Chapman Fuqua Conservatory at Atlanta Botanical Garden, **40–41**
Downing, Jason, 197
downy rattlesnake plantain (*Galearis spectabilis*), 243
Dracula, 39, 214, 216
  as pleurothallids, 39
  D. inexperata, 94
  D. lemurella (ghost monkey orchid), 210, **211**
  D. sodiroi, 214
Drakaea, 32, 83
  D. livida (ecotypes), 34
Dressler, Robert L., 34, 288
Droissart, Vincent, 28, 277
drones, 167
drought(s), 39, 129, 134, 140, 143, 144, 152, 155, 158, 238, 239, 252
Dryden, Kath, 237
Dublin, Ireland, 181
Duke of Devonshire, 56
dunes, 192, 193
Dunsterville, 'Stalky', 63
Duperrex, Aloys, 283
dust mites (*Dermatophagoides*), 274

E

earthquake(s), 74, 279
echidna, 83
ecological succession, 185, 208, 222
EcoMinga Foundation, 206, 215, 218
ecosystem, 8, 87, 159, 164, 190, 193, 208, 209, 210, 217, 227, 259, 288
ecotypes, 34, 222
ecotourism, 49, 82, 94, 95, 117, 209, 210, 216, 288
Ecuador, 63, 72, 129, 130, 134, 136, 138, 206, 213, 214, 215, 216, 217, 234, 265, 272, 276, 278, 279, 280, 281, 285, 288
  Antisana National Park, 280
  Baeza, 280
  Baños, 94, 215
  Cuenca, University of, 95, 279
  Dracula Reserve, 216, 218
  El Pahuma Orchid Reserve, 213, 214, 217
  Guayaquil, 272
  Jardin Botánico de Quito, 265, 276
  Loja, 94
  Oro, 94
  Papallacta, 138, 280
  Quito, 136, 138, 214, 280
Edo period, 113
El Niño, 158
Elleanthus lupinus, 33
Ellyard, R. K., 86
Elórtegui Francioli, Sergio, 11, 192
Encyclia, 176
  E. brevifolia, 106
  E. citrina, 99
  E. phoenicea (chocolate orchid), 106
endemics, 48, 49, 63, 66, 67, 69, 70, 72, 73, 75, 79, 87, 94, 95, 106, 130, 136, 138, 139, 140, 142, 143, 171, 175, 180, 189, 190,

192, 207, 211, 218, 231, 250
English Channel, 154
English lady's-slippers, 235–239
English meadows and orchards, 244–246, **247**
Epidendrum montserratense, 140
Epigeneium fargesii, 170
Epipactis thunbergii, 32
epiphyte, 35, 38, 39, 71, 75, 83, 98, 113, 135, 145, 158, 166, 167, 170, 172, 176, 182, 197, 200, 207, 208, 213, 222, 223, 225, 252, 288
 twig, 42
Epipogium aphyllum, 98
Eria rhomboidalis, 131
Eric Young Orchid Foundation (EYOF), 260
Erickson, Rica, 84
Estonia, 276
Eulophia, 116
 E. graminea 36
Europe, 9, 35, 48, 51, 55, 56, 57, 58, 59, 62, 104, 116, 126, 144, 149, 151, 154, 185, 210, 230, 235, 237, 238, 241, 278, 281, 283, 284
Everglades, 9, 12–13, 14, 15, 159, 223, 291, 292
evolution, 9, 28, 37, 38, 62, 66, 74, 99, 101, 177
ex situ, 87, 91, 116, 151, 162, 225, 231, 233, 237, 241, 256, 259, 265, 292
extinction, 8, 10, 11, 21, 30, 66, 69, 79, 84, 86, 119, 124, 125, 126, 129, 138, 140, 145, 149, 150, 155, 171, 189, 217, 230, 234, 279, 289, 292
extirpation, 116, 124, 125

F
Fairchild Tropical Botanic Garden, 197, 291
 'mobile orchid lab', **198**
Fast, Gertrud, 283
Fay, Mike, 230
fermentation, 31
ferns, 39, 63, 101, 152, 170, 176, 192, 207, 208, 252, 253
 Blechnum, 158
 Davallia clarkei, 170
 Osmunda regalis (royal fern), 56
 Phymatopteris quasidivaricata, 170
 Rhododendron kawakamii, 170
Ferreira, Danielle, 225
 pollinating *Cyrtopodium punctatum*, **227**
fertilizer(s), 246
fire (burning), 53, 72, 80, 90, 91, 94, 107, 124, 126, 131, 134, 135, 143, 152, 155, 156, 158, 175, 176, 233, 234
fitness (ecological), 222, 238
Flanagan, Mark, 131
flasks, 58, 147, 151, 164, 181, **183**, 198, 199, 291
 germination studies, **253**, 274,
 ghost orchid seedlings, 291
 Erlenmeyer flask, 283
flavouring, 117, 119
flies, 30, **211**
floods, 21, 134, 155, 156, 185, 208, 209, 234, 268
Florida, 9, 14, 15, 17, 18, 19, 20, 21, 35, 36, 37, 42, 98, 125, 138, 155, 159, 183, 223, 225, 226, 227, 231, 291, 292
 Big Cypress Swamp, 14
 Everglades, 9, **12–13**, 14, 15, 159, 223,

291, 292
Fairchild Tropical Botanic Garden, 197, 198, 291
Fakahatchee Strand, 15, 17, 18, 21, 223, 225, **226**
Florida Panther National Wildlife Refuge, 18, 20, 223
Ft. Myers-Naples, 15
Miami, 15, 197, 198, 199, 227, 291
Naples Botanical Garden, 24, 147
Tamiami Trail, 15
University of Florida, 23, 24
forest (see also, cloud forest),
 Amazon, 15, 135, 170, 171
 Atlantic, 170, 171, 172, 175, 177, 199, 207, 222
 Araucaria mixed, 171
 climax (old growth), 136, 167, 222
 dwarf montane, 95
 elfin, 134
 heathland, 87
 lowland, 63, 251
 moist, 11, 83, 134, 145, 172
 monsoon, 63
 montane, 156, 158, 172, 201, 213, 259
  dwarf, 95
  evergreen, 94
 Polylepis, 280
 primary, 70, 134, 167, 216, 222, 251
 remnant in Ecuador, **136**
 riparian, 199
 sclerophyllous, 192
 secondary, 70, 176, 211, 222
 semi-deciduous, 171, 189, 199
 tropical dry, 63, 259
 tropical rainforest, 28, 62
 thorn, 63
fragrance (perfume), 30, 31, 113, 114, 117, 119, 200, 208, 256, 259
France, 124
 Alsace, 235
 Boissy-Saint-Léger, 260
Franklin, Sir John, 152
frass, 39, 273
Freestone, Marc, 87
fresco, 80
frogs, 8
 Hyloscirtus conscientia, 218
 Pristimantis species, 218
Frosch, Werner, 283
frosts, 52, 72, 134, 199, 256
fungus (fungi) (see also, mycorrhizal fungi),
 fungus gnat, 30, 32
 Fusarium wilt, 119
 mould, 273

G
Galapagos Islands, 67
Gale, Stephan, 145
gambir plantations (*Uncaria gambir*), 251
Gardens by the Bay, Singapore, 190, **191**, 192
Gardner, George, 107, 108
gas chromatography, 30
Gastrodia elata, 116
Gates, Bill and Melinda, 244
gene bank, 223
gene pool, 240
genes, 222, 238
genetic, 101
 blueprint, 34

code, 17
diversity (variation), 28, 87, 119, 225, 230, 233, 238, 240, 244, 253, 256, 257, 258, 259, 264, 265, 268, 271, 273, 278
fingerprinting, 91, 245
material for hybrids, 113
resources, 151
vigour, 240
genets, 231, 233
genus, 98, 101
Germany, 57
germination, 20, 23, 35, 37, 38, 82, 120, 144, 180, 193, 194, 198, 200, 222, 225, 237, 270, 271, 275, 276, 277, 283, 284
 asymbiotic, 120, 182, 198, 237, 285
 germination testing with tetrazolium, 198, **266–267**, 276
 symbiotic, 23, 85, 86, 120, 180, 198, 222, 237, 238
germplasm, 198, 241
ghost orchids, 9, 15, **16**, 17, 19, **20–21**, 22, 23, 24, **25**, **31**, **32**, 37, 98, 107, 138, **139**, 147, 151, 155, 197, 231, 291, **292**, **293**
 Dendrophylax lindenii, **14**, 15, 18, 19, **20–21**, 23, **25**, **31**, **32**, **106**, 125
 Didymoplexis stella-silvae, 98
 Epipogium aphyllum, 98
  nectar of, 31
Gigot, Guillaume, 102
glaciers, 154, 155
Glasgow Botanic Gardens, 107
glasshouse (see greenhouse)
global warming (see climate change)
gold, 134, 216
Grammatophyllum speciosum ('tiger orchid'), 35, **36**, **251**, 253
Grand Cayman, 98, 138, 180
grandchildren, 209, 250, 289
Grant, Ruth, 200, 201
grassland(s), 72, 87, 90, 91, 153, 244, 246, 249
green pod, 120
greenhouse, 23, 24, 56, 57, 58, 59, 84, 100, 107, 113, 151, 159, 162, 198, 199, 234, 240, 274, 290
greenhouse gas emissions, 152
Grey, Robert, 49
Guadalupe, school project, **197**
Guarianthe skinneri, 112
Guatemala, 28, 55, 114, 144, 145, 230, 276
guava, 159
Guidoni Maragni, Angélica, 222
Gulf Stream, 154
Guterres, Antonio, 6
Gymnadenia, 283
 G. conopsea (fragrant orchid), 245
gymnosperms, 171

H
Habenaria repens, 183
Halbinger, Federico, 163
hammer orchids (*Drakaea* species), 32
hammock, 223
Hargreaves, Serene, 120
Hawaii, 36, 66, 67, 68, 69, 74, 98, 159, 182, 234
 Alakai Swamp, **68**
 Kauai, 67, 68, 69
 Lyon Arboretum, 69
 Maui, 68

298   SAVING ORCHIDS

Molokai, 68
Mauna Kea, 67
Oahu, 68
The Big Island, 67
University of Hawaii, 69
Heathrow Airport, 151
heat island, 140, 249
heat waves, 152, 156
Helmich, Rachel, 271, 272
Hennis, Wilhelm, 53, 55
herbarium, 55, 70, 73, 84, 98, 100, 102, 103, 120, 193, 234, 290
 school herbarium in Chile, **194**
 specimens 57, 72, 103, **104**, 144, 235, 263, 273
Herdman, Adam, 19
Hermans, Johan, 94
Hetaeria oblongifolia (rediscovery), 95
Hoang, Nguyen, 23
Holland (see Netherlands)
Honduras, 114, 230
honey fungus (*Armillaria mellea*), 116
Hooker, Sir William Jackson, 57, 105
Howarth, Francis and Mull, William, 67
Hsu, Rebecca, 11, 129, 167
Huanglong, China, **92–93**, 94, 237, 279
 Huanglong Nature Reserve, 243
 Huanglong Valley, 280
Humblot, Léon, **46–47**, 48, 49
Humboldt, Alexander von, 8, 49, 66, 280
humidity, 166, 176, 190, 199, 207, 225, 231, 250, 252, 256, 270, 279, 280, 281
 dome, 20, 23
 humidifiers, 192
 in cloud forests, 217
 relative, 57, 252, 268, 274
hummingbirds, 32, 49, 83, 106, 141, 143, 176
 bee, 106
 marvellous spatuletail (*Loddigesia mirabilis*), 49
humus, 106, 158, 166, 176, 177, 243, 246, 249
Hungary, 239
 Bükk National Park, 239
Hutchens, Johanna, 151
Hutchings, Michael, 153, 185
hybrid(s), 36, 58, 113, 198, 231, 256, 260, 264, 284
 natural, 94, 241
 swarms, 99
hypha(e), 22

I
ice sheets (see glaciers)
illegal collection (poaching), 17, 21, 100, 117, 129, 131, 134, 144, 145, 163, 172, 210, 240, 290
 at Kings Park and Botanic Garden, 146, 147
 of *Laelia speciosa*, **150**, 163
 of *Paphiopedilum helenae*, 230
Illinois, 19, 68, 227
 Chicago Botanic Garden, 151, 291
 Illinois College, 77, 79, 271
 Jacksonville, 271
inbreeding depression, 240
in vitro culture, 143, 151, 175, 213, 237
India, 32, 75, 114, 119, 126, 142, 276
 Kerala, 142
 Punjab, 62
indigenous people(s), 98, 114, 159, 163, 216

Guarani, 175
Indonesia, 66, 69, 70, 73, 134, 276, 279
 Bali, **70–71**
 Bogor Botanic Gardens, 70, 72
 Cibodas Botanic Garden, 70, 72
 Indonesian Institute of Science, 70
 Java, 35, 70
 Kalimantan, 70
 National Research and Innovation Agency, 70
 Sulawesi, 70
 Sumatra, 35, 70
 Sunda group, 66
inselberg, **132**
Inter-American Highway (see Pan American Highway), 288
International Orchid Conservation Congress (IOCC), 86, 235
international trade, 145, 147, 149, 150, 151, 216, 230, 265
International Union Conservation of Nature (IUCN), 90, 125, 126, 139, 223, 235
 Orchid Specialist Group, 230, 289
internet, 23, 84, 147, 188, 231, 290
invasive species, 134, 158, 159, 172, 190, 227, 235
 inchworm caterpillars (*Eupithecia*), 66
 invasive orchids, 37
 Oeceoclades maculata, 159
 Phaius tankervilleae, 159
 Polystachya concreta, 159
Ionopsis utricularioides, 42
Iran, 119
Isotria medeoloides, 182

J
Jamaica, 8
Jameson, 108
Janzen, Dan, 99
Japan, 32, 113, 116, 149, 154, 238, 241
Jaramilla, Paola, **265**
jipijapas, 48
Johanssen, Carl, 48, 52
Johnson, Lynnaun, 22, 23
Jost, Lou, 94, 215

K
Kane, Mike, 23
kangaroo(s), 50, 83
 tree, 74
kangaroo grass (*Themeda triandra*), 90
karst limestone, 19, 72
karyotype, 102
Kefersteinia retanae, 126, **128**, 129
Kell, Shelagh, 235
Kendon, Jonathan, 79, 82, 120
Kenya, 277
Kesseler, Rob, 268
Kew Micropropagation Unit, 237
Kienast-Zölly, Ludwig, 164
Kilner jar, 270, 274
Kirkham, Tony, 131
Knight nursery, 105
Knudson, Lewis, 182
koala, 83
Koehler, Samantha, 223
Kolanowska, Marta, 95
Komodo dragon (*Varanus komodoensis*), 66
Kondrat, Justin, 259
Koopowitz, Harold, 124

Korea, 113
Krakatoa, 35

L
labellum (lip), 30, 32, 107, 138, 230
laboratory, propagation, 23, 82, 87, 143, 149, 182, 198, 199, 213, 225, 251, 292
 at Atlanta Botanical Garden, **275**
 in China, **117**
Lady's Slipper Orchid Committee, 234
Laelia, 114
 L. anceps, **29**, 58, 62, **63**, 162, 163, 164, **258**
 L. anceps var. *dawsonii*, 58, 95, 163, 182
 L. anceps var. *mayensis*, 162
 L. crispa, 176
 L. dawsonii, 162,163, 164
  f. *chilapensis*, 162, 163, **165**, 216
  f. *dawsonii*, 163, 164, **165**
 L. gouldiana, 125
 L. lobata, 108
 L. speciosa, 149, **150**
  painting of, **148**
 L. superbiens, 58
Lager, John, 48, 50, 53, 256, 259
laminar flow hood, 274, 275
landslide(s), 167
Laos, 197
Latin names, 98
Latin America, 50, 189, 276, 277, 278
leafless orchid(s), **14**, 37, 42, 82, 98, 138, 292
Lecoufle, Marcel, 124, 260, 261
lemurs, 48, 63, 75, 82
Lepanthes, 35, 39, 94, 145, 208, 215, 218, 250, 280, 281
 as *pleurothallids*, 39
 L. caranqui, 94
 L. caritensis, 207
 L. dougdarlingii, **28**
 L. microprosartima, 94
 L. orolojaensis, 94
 L. tibouchinicola, 207, **212**
 L. tulcanensis, 218
 L. wakemaniae, **218**
lichens, 22, 39, 63, 170, 180, 207, 208, 260
Lieberman, Diana, 134
Ligon, Jason, **275**
lightning, 19, 39, 170
lily of the valley (*Convallaria majalis*), 113
Lima family, 215
limestone, 19, 72, 155, 230, 231, 246, 280
Linden, Lucien, 48, 105, 181
Lindley herbarium, 98, 103
Lindley, John, 57, 103, 104, 144, 162
Linnaeus, Carolus, 98
Listera ovata (twayblade), 185, 245
lithium chloride, 274
lithophyte (lithophytic), 39, **78**, 81, 82, 113, **133**, 158, 175, 199
 Aerangis ellisii, 81, 82
 Angraecum longicalcar, **78**, 82
liverworts, 180
livestock (see cattle)
living collection(s), 11, 100, 102, 199, **254–255**, 256, 257, 261, 263, 264, 273, 277, 290
lizards, 66, 218, 253
Loddiges, George, 49, 56, 57, 107, 144, 162, 264
 hummingbird collection, 49
logging, 124, 167, 222, 223

London Zoo, 48
lost species, 129
Lucas, Gren, 234
Luer, Carlyle A., 17, 68, 69
lumber, 15, 74, 80, 135
Luo, Youqing, 257
    on scarab beetle, 257
Lycaste skinneri, 144

M
McCormick, Melissa, 243
McGuinness, Julianne, 244
McIllmurray, Michael, 263, 290
Macaé de Cima, Brazil, 171, 175, 176
Machado-Neto, Nelson, 199, 200, **201**, 270, 272
machete, 53, 55, 82
Madagascar, 28, 31, 36–47, 48, 49, 63, 75, 77, **78**, 80, **81**, 82, 94, 98, 117, 119, 131, **132**, 180, 181, 210, 234, 260, 278
    Ambatofinandrahana, 79
    Analabeby, 79
    Antananarivo, 75, 80, 278
    Central Highlands, 75, **81**
    Itremo Plateau, 79
    Kew Madagascar Conservation Centre, 75, 234
    Mahavanona, 82
    Mihary Soa Preserve, 80, 131
    mural, **78**
    Tsimbazaza Botanical and Zoological Park, 80
magnolias, 208
malaria, 49, 66, 81
    avian, 66
Malay Archipelago, 70
Malaysia, 134
    Sabah, 201
Malmgren, Svante, 237, 283, 284
mammals, 32, 83, 112, 206, 208, 209, 217
    agouti, 42
    Andean spectacled bears (*Tremarctos ornatus*), 217
    brocket deer (*Mazama* sp.), 217
    giant panda (*Ailuropoda melanoleuca*), 217, 234
    mice,
        Pattonimus ecominga, 218
        Chilomys georgeledecii, 218
    ocelot (*Leopardus pardalis*), 217
    pacaranas (*Dinomys branickii*), 217
    platypus, 83
    puma (*Felis concolor*), 217
    shrew opossums, 218
Mandela, Nelson, 188
Maryland, 243
mass extinction, 289
mass spectrometry (see gas chromatography)
Masters, Susanne, 129
Masdevallia, 39, 63, 138, 250
    as pleurothallids, 39
    M. amanda, 213
    M. hortensis, 213
    M. macrura, 208
    M. rimarima-alba, 116
    M. rosea, 138
    M. uniflora, 116
    M. ventricularia, 208
Materials Transfer Agreement, 279

Mathers Foundation, 260
Maxillaria, 263, 290
    M. tenuifolia, 30
Mead, Margaret, 288
meadow, 158, 185, 245, 246
    meadow saffron (*Colchicum autumnale*), 245
    English, 42, 185, 244, 245
    hay, 246
mealybugs, 159
Medellín, 50, 119, 183, 189, 197
    in 1948, **248**, 249, **250**
media (agar), 22, 91, 151, 182, 196, 198, 225, 237, 264, 274, 276, 283
    Knudson C, 277
    Malmgren's, 283, 284
    TGZ, 283
medicine (medicinal), 145, 188
    alternative, 17
    medicinal orchids, 98, 114, 116, **117**
    traditional Chinese, 114, 116, 217
Mediterranean, 42, 119, 154, 156, 190, 192, 284
Melaleuca, 83
Melastomataceae, 176
Merchán, Francisco, **265**
mercuric iodide, 58
meristemming, 264, **265**
Mesoamerica, 116, 206
Mexico, 28, 62, 63, 113, 114, 117, 124, 143, 144, 149, 152, 156, 159, 162, 163, 164,166, 216, 230, 231
    Alto Lucero Veracruz, 62
    Chiapas, 162
    Guerrero, 163
    Gulf of Mexico, 164
    Jalisco, 95, 163
    Los Chimalapas, 231
    Mexico City, 50, 103, 142, 143, 149
    Michoacán, 149
    Montebello, 134
    Oaxaca (Oajaca), 114, 162, 163, 231
    Reserva Pedregal de San Angel, 142
    Universidad Nacional Autonoma de Mexico (UNAM), 143
    Universidad Veracruzana, 166
    Veracruz, 149, 156, 162, 163, 164
Mexipedium xerophyticum, 95, 125, 231, **232**, **236**, 263
Micholitz, Wilhelm, 48
microbes (microorganisms), 23, 31, 140, 154, 156
micropipette, 32
microscope, 101, 196, 275
    dissecting, 35, 38
microscopist's thumb, 101–102
mignonette leek orchid (*Prasophyllum morganii*), 124
Millennium Seed Bank (Kew), 35, 140, 237, 257, 268, **269**, 274, 276, 277, **278**, 279, 290
    Partnership, 277
Miller, David, 135, 171, 175, 176, 177, 207
Miller, Izabel 171, 177, 207
Miller, Scott, 134
Macaé de Cima, 171, 175 , 176
Millican, Albert, 48, 50, 52, 53, 59
Million Orchid Project, 197, **198**
Milne-Redhead, Edgar, 234
Miltoniopsis,
    M. roezlii, 50

M. vexillaria, 50, 59, 198, 208, 213
minerals, 37, 182, 246, 279, 284
mining, 75, 79, 90, 134, 216
Missouri Botanical Garden, 272
Mitchell, Robert, 237
mites, 22, 274
mobile orchid lab, 197, 198
moisture (see humidity)
molecular genetics, 99
monitor lizard (see Komodo dragon)
monoculture, 119, 259
Moore Ltd (orchid nursery), **54**
Morales, Juan, 164
Morris, Daniel, 8
mosquito(es), 19, 32, 49, 66, 106
    net, 81
moss(es), 39, 56, 63, 72, 176, 177, 180, 207, 208, 260, 291, 292
    clubmosses, 192, 280
moths, 21, 30, 31, 32, 69, 100, 272, 291
    sphinx (hawk) moths, **20**, **21**
mountain(s),
    Andes, 48, 105, 131, 189, 206, 208, 209, 210, 213, 215, 249, 250, 280, 281
        Central Cordillera, 105, 208, 250
    Cerro Candelaria, 215
    Cordillera de la Costa, 189
    Cordillera de Talamanca, 216
    Himalayas, 32, 75
    Hengduan Mountain Region, 279
    Jura, 235, 240, 246
    Kinabalu, 201, 210
    Min Shan Range, 279
    Organ Range, 107, 175
    Serra da Cantareira, 172
    Swiss Alps, 240
    Xuebaoding, 279
mowing, 140, 249
Mozambique Channel, 75, 80
Mújica, Ernesto, **18**, 19, 21
mulch, 86
mules, 55, 259
mushrooms, 8, 211, 285
Myanmar, 54, 233
mycology, 85
mycorrhizal fungi, 20, 21, 23, 37, 38, 82, 83, 85, 86, 91, 98, 100, 101, 120, 143, 154, 156, 181, 182, 193, 196, 208, 222, 234, 235, 237, 243, 244, 250, 271, 285
    storage of, 285
mycotrophy, 42, 182
Myrmecophila thomsoniana (banana orchid), 180

N
Naples Orchid Society, 19
National Aeronautics and Space Administration (NASA), 155
National Orchid Conservation Centre (China), 233
National Plant Collections, 261, 263
natural selection, 28, 62
Nature's Calendar, 153
nectar, 28, 30, 31, 42, 192, 249
    reward, 28, 69
    spur, 32, **33**, 63, **77**, **106**
nectaries, 113
Neofinetia (*Vanda*) *falcata*, 113, **114**
Neotinea ustulata (burnt orchid), 153
Neottia nidus-avis (bird's-nest orchid), 42, **43**, 182

Nepal, 116
　Annapurna Research Centre Kathmandu, 117
Netherlands (Holland), 113, 155, 244
　Utrecht University, 244
Neuwiedia nipponica, 34
New Guinea, 36, 70, 72, 74, 180, 234
　Tari Gap, 72, 73
New World, 35, 39, 63, 207
New York, 288
New Zealand, 107
niche(s), 39, 66, 67, 125
nitrogen, 39, 245, 277, 283
North America, 9, 55, 66, 90, 152, 153, 156, 182, 230, 234, 241, 244, 269, 279, 291
North American Orchid Conservation Center (NAOCC), 243, 244, 279, 291
North Pole, 152
Northern Hemisphere, 22, 24
Northwest Passage, 152
nursery(-ies), 48, 50, **54**, 55, 56, 57, 59, 74, 87, 91, 113, 124, 126, 140, 146, 147, 149, 181, 199, 207, 208, 211, 216, 231, 234, 241, 243, 260, 264
nutrients (see minerals)

O

ocean plate, 66
Odontoglossum, 181
　O. bluntii (syn. *O. crispum* and *Oncidium alexandrae*), 50, **51**, 52
　O. crispum (syn. *O. alexandrae*), 2, 9, **10–11**, **51**, 52, 53, 55, 58, 59, 98, 99, 163, 181
　　'Avalanche', 260, **262**
　　branched specimen, **99**
　　collecting in Colombia (painting), **44–5**
　　herbarium specimen, **96**
　O. hallii, 214
　O. mirandum, 213
　O. sceptrum, 213
Old World, 72, 119
Oncidium, 176, 280
　O. alexandrae, **11**, **98**, 99
　O. flexuosum, 136
　O. hians, 223
　O. kramerianum (syn. *Psychopsis krameriana*), 50
　O. varicosum, 172, **173**
On the Origin of Species, 28
Ophrys, 235, 283
　as an aphrodisiac, 116
　bee orchid (*O. apifera*), 140, 153, 185, 245, 246, 249
　early spider orchid (*O. sphegodes*), 36, 153, 185
　O. exaltata subsp. *morisii* in Sardinia, 140
　O. insectifera, 140
　O. tenthredinifera, **26**, 154
Opuntia streptacantha, 143
Orchis maculata, 34
orchards, 35, 86, 244, 245, 246
Orchid Conservation Chelsea, 291
Orchid-Gami, 291
Orchid Kit, 86
Orchid Review, 105, 135, 145, 181, 280
Orchid Seed Bank Challenge, 277
Orchid Seed Stores for Sustainable Use (OSSSU), 118, 200, 270, 276, 277, 279
Orchids magazine, 129

organic matter (see humus)
Orlean, Susan, 17
ornithologist(s), 49, 213, 217
Ortega-Larrocea, Pilar, 143, 144
osmunda fibre (*Osmunda regalis*), 56
Otway, Beth, 290
outbreeding depression, 238
Oversluys, (his forename is lost in the mists of time), 48, 55
ovary, 34, 42
oxen, 55
oxygen, 250, 279

P

Pacific Ocean, 66, 69, 74, 105, 155, 234
Pacific Plate, 66
Pakistan, 119
Palau (Micronesia), 74, 75, 155, 234
palm house, 58
palm oil plantations, 134
Panama, 163, 276
　Panama Canal, 163
Pan-American Highway, 138
Panisea cavaleriei, 131
Pant, Bijaya, 116
papaya, 159
Paphiopedilum, 72, 84, 114, 230
　P. armeniacum, 233
　P. dayanum, 201, **202**
　P. delenatii, 124
　P. druryi (India's 'golden paph'), **141**, 142
　P. fairrieanum, 126, **127**, 129
　P. haynaldianum var. *album*, 263
　P. helenae, 230
　P. hirsutissimum, **130**, 131
　P. insigne, **112**, 113
　P. nataschae, 146
　P. philippinense, 101, **102**
　P. rothschildianum, 201, **202**
Papua New Guinea, 72
Paracaleana disjuncta, 90
Paraguay, 171
páramo, 95, 209, 280
Paraphalaenopsis serpentilingua, 70
Parke Wright IV, John, 291
Patagonia, 180
Pate, John, 85
Pauwels, Theodore, 48, 50, 54, 59
Paxton, Joseph, 107
　Paxton's Flower Garden, 105
pedregales (old lava fields), 143, 280
Perdomo, Vicente, 260
perfumes (see fragrance)
Peristylus holochila (syn. *Platanthera holochila*), Hawaiian bog orchid, 68, 69, 98, 182
Perkins, Fern, 152
Perlman, Steve, **68**
Permian, 171
Perner, Holger, 156, 158
Peru, 49, 63, 94, 116, 206, 214, 231, 234
　central highlands, 116
　Huembo Reserve, 49
pests, 151, 159, 222, 225, 227, 256, 257
　mealybug (*Pseudococcus microcirculus*), 72, 159
Petri dish(es), 22, 272, 274
Peyrot, Jean-Pierre, 260, 261
pH, 243
phytosanitary certification, 147, 150, 151, 291

Phalaenopsis, 290
　hybrids, 113
　P. cornu-cervi, **220**
　P. amabilis, 58, 163
　P. grandiflora, 163
　P. serpentilingua, 70
pharmacies, 188
phenol, 56
phenology, 213
pheromones, 34
Philippines, 73, 74, 276
　National Herbarium, 73
Phragmipedium, 230, 231
　P. besseae, 94, 230, 231
　P. exstaminodium, 125
　P. kovachii, 94, 231
photosynthesis, 37, 38, 39, 42, 182
pines, 63, 190, 246
　Pinus radiata, 94
pipers, 207
Pizarro, Gonzalo, 280
plant blindness, 188
plant hunters, 48
　Humblot, Léon, **46–47**, 48, 49
　Johanssen, Carl, 48, 52
　Lager, John, 48, 50, 53, 256, 259
　Linden, Lucien, 48, 105, 181
　Micholitz, Wilhelm, 48
　Millican, Albert, 48, 50, 52, 53, 59
　Oversluys, 48, 55
　Pauwels, Theodore, 48, 50, 54, 59
　Roezl, Benedikt, 48, 105
　Skinner, George Ure, 48, 49, 144
　Wallis, Gustave, 48, 50, 105
　Wilson, Ernest H., 48, 131, 279, 280
Platanthera, 90
　P. bifolia (lesser butterfly orchid), 153
　P. chlorantha (butterfly orchid), 246
　P. holochila, 68
　P. hyperborea var. *viridiflora*, 68
　P. integrilabia, 271, **272**
　P. leucophaea, 182
　　seed, **269**
　P. praeclara, 182
Platystele, 215
Pleione formosana, **169**
Plectorrhiza purpurata, 158
Pleurothallids, 32, 35, 39, 124, 144, 145, 183, 215, 261
Pleurothallis, 218
　as pleurothallids, 39
　P. acinaciformis, 208
　P. chicalensis, 218
Poilane, 124
pollen, 21, 28, 30, 31, 34, 151, 200, 208, 230, 240, 249, 257
　exchange, 257
pollination, 9, 21, 28, 30, 31, 34, 42, 166, 193
　artificial (hand), 119, 120, 166, 176, 233, 237, **239**, 240, 272, 290
　cross, 194, 257
　natural, 177, 237, 238, 252
　self-pollination, 273
pollinators, 21, 28, 30, 31, 32, 34, 86, 87, 106, 119, 125, 134, 136, 153, 154, 166, 172, **173**, 175, 192, 193, **203**, 211, 250, 252, 281, 291
　bees, 30, 135, 172, 192, 200, 208, 249
　　Braunsapis clarihirta, 253
　bumblebees, 176
　　Bombus, 175

carpenter bees, 176
euglossine bees, 42, 176, 200, **203**
    Eulaema, 166
Megachile bees, 253
solitary bees (*Andrena nigroaenea*), 153
beetles, 208
birds, 32
    hummingbirds, 32, 143, 176
    sunbirds, 32
butterflies, 30, 272
    mountain pride butterfly (*Aeropetes tulbaghia*), 32, 33
decline in, 172
moths, 21, 31, 272
    giant sphinx moth (*Cocytius anteaus*), 21
    hawk moth (*Pachylia ficus*), **20**, 21
mosquitoes, 32
mountain mouse (*Rattus fulvescens*), 32
raspy cricket (*Glomeremus*), 32
wasps, 34, 83, 172
pollinia (-ium), **20**, 21, 28, 30, 32, **33**, 34, 100, 150, 192, 194, 200
Polynesians, 98
Pope Francis, 75
population studies (surveys), 42, 87, 100, 153, 211, 241
    of European terrestrial orchids, 185
Posada, Juan Felipe, 135
Posada Ochoa, Mario, 198
posters,
    Paphiopedilum rothschildianum (Ruth Grant), **202**
    Coryanthes kaiseriana (Ian Cartwright), **203**
pothole, 79, 80
Pradhan, Udai, 127
Pradhan, Hemlata, 127
prairie, 90
primates,
    gibbons, 63
    monkeys, 49, 63
    orang-utans, 63, 70, 74
primitive angiosperms, 208
Pritchard, Hugh, 234, 276
proboscis, **20**, 21, 22
Prosthechea,
    P. citrina (syn. *Cattleya citrina* and *Encyclia citrina*), 99
    P. cochleata, **27**
    P. michuacana, 114
    P. karwinskii, **115**, 116
    P. vitellina, rescue, **122–123**, **166**
protein, 208, 283
Protheroe and Morris (auctioneers), 52
protocorm(s), 37, 38, 67, 82, 196, 272, 284
Pseudolepanthes bihuae, 218
Peristylus holochila, 98, 182
Pterostylis,
    P. aenigma, 90
    P. vittata (syn. *P. sanguinea*) (banded greenhood), 84, 85
pseudobulbs, 39, 63, 68, 72, 114, 116, 144, 209, 225, 252
pseudocopulation, 34, 82
Pterostemma antioquiense, 213
Puerto Rico, 207
    Carite State Forest, 207
Pugh-Jones, Simon, 196
Puspitaningtyas, Dwi Murti, **70**, **71**
Pupulin, Franco, 94, 124
putty root (*Aplectrum hyemale*), 243

Q
quarantine(s), 222

R
Radcliffe, Ellen, 23
Raffles, Sir Stamford, 251
Rafflesia, 63, 69
rainfall, 55, 63, 152, 158, 209
Ramos, Andres, 164
Ramos, Augustin, **122–123**
Ramsay, Margaret, 237
Ramsbottom, John, 181
Rasmussen, Hanne, 38, 235, 243
Rajaovonirina, Landy, 75, 79
Ratovonirina, Gaëtan, 79
Raven, Peter, 134
Red Listing, 125, 183
    of Threatened Species, 90, 125
    Global Red List, 125, 126
reef triggerfish (*Rhinecanthus rectangulus*), 98
Reichenbach, Heinrich Gustav, 57
reintroduction, 11, 23, 83, 91, 143, 144, 147, 182, 196, 199, 200, 222, 223, 235, 237, 238, 252, 253, 257, 272, 277
    in Brazil, 222
    of *Bletia urbana*, 143, 145
    of *Cattleya purpurata* (donkey's ear), 175
    of cigar orchid (*Cyrtopodium punctatum*) in Florida, 223
    of *Cypripedium calceolus*, in the UK, 235, **236**, 237
    in Switzerland, 145, 240, **241**, **242**
    of *Cypripedium macranthos* in China, 240, 241, 243
    of orchids in Singapore, 222, 253
    of *Paphiopedilum armeniacum* (golden slipper) in China, 233
Reiter, Noushka, 86, 87
relative humidity (see humidity)
Reserva Natural La Selva de Ventanas, 211
resin, 53
Restrepia (as pleurothallids), 39
Restrepo, Catalina, **183**
Reunion Island, 32, 119
Rhizanthella ('underground orchids'), 83
    R. gardneri, 83, 84
    R. johnstonii, **85**
    R. slateri, 83
rhizomes, 79, 95
Ridley, Henry Nicholas, 95
Rio Atlantic Forest Trust (RAFT), 171
Rittirat, Suphat, 35
River of Grass, 15
roadside verge, 249
Roberts, David, 79, **132**
rodents, 34, 42, 217
    rats, 57
Roezl, Benedikt, 46, 105
Rolfe, Robert Allen, 105
Rollison, 107
Romania, 145
roots, **15**, 17, 19, 22, 23, 24, 37, 38, 39, 42, 56, 57, 72, 79, 82, 100, 156, 177, 181, 208, 222, 225, 237, 245, 280, 292
Rothschild, Lionel, 58
Royal Botanic Gardens, Kew, 86, 94, 140, 142, 177, 230, 234, 237, 257, 263, 268, 269, 276, 278, 285, 290, 291, 292
    Jodrell Lab, 101
    Lindley herbarium, 98, 103
    Living Collection, 102
    orchid herbarium, 102, 104
    Princess of Wales Conservatory, 292
    Wakehurst, 35, 234, 237
Royal Horticultural Society (RHS), 201, 263
Rucker, Sigismund, 103, 104
Russia, Leningrad (St. Petersburg), 268
Rwanda, 197

S
Sagan, Carl, 271
Sainsbury Orchid Conservation Project, 75
Sainsbury, Sir Robert and Lady, 237
Samala, Sainiya, 35
samurai warriors, 113
Sánchez-Cuervo, Ana María, 209
Sander, David, 182
Sander, Frederick, 48, 50, 55, 57, 72, 126, 164, 234
Santuario de Los Molles, Chile, **188–189**
Sarasan, Viswambharan, 75, 79, 80, 82, 142
satinwood, 58
savannah, 170, 199
scale insects, 159, 225
Scandinavia, 238
Scaphosepalum zieglerae, 218
Schettler, Roland, 72
Schimper, A. F. W., 62
school in Guadalupe, Colombia, **197**
schoolchildren (see children)
Schröeder, Baron Bruno, 58
Schuiteman, André, 103
Scientific Orchid Conference on Andean Orchids, 197
scorpions, 55, 57
sea level rise, 9, 154, 155
Searight, 126
season,
    dry, 56, 63, 72, 199
    rainy (wet), 21, 79, 80, 81, 82, 199, 222, 243
Seaton, Philip (Phil), 79, 101–102, 106, **112**, 120, 234, 239
seed,
    baiting, **77**, 144
    bank, 11, 87, 139, 151, 158, 198, 268, 273, 274, 276, 277, 278, 279 (see also, Millennium Seed Bank (Kew)
        in soil, 182
        Kunming Institute of Botany, 277
        UNOESTE, 200
    capsules (see capsules)
        cleaning, drying and storing, 274–275
    coat (testa), 34
    collection, 120, 225, 273
    counting, 35
    dispersal, 35, 74, 83, 250
    dormancy, 182, 237
    embryo(s), 11, 34, 37, 143, 180, 181, 237, 252, 266–267, 273, 274, 275, 276
    growing from seed, 23, 25, 68, 74, 149, 181, 182, 183, 211, 222, 225, 233, 234, 237, 260, 261, 272, 283
        Acineta barkeri, 166
        Angraecum longicalcar, 82
        Caladenia cretacea, **88–89**
        Caladenia cruciformis, **88–89**
        Cypripedium calceolus (Lady's slipper orchid), 234
        Dendrobium officinale, 116
        Laelia dawsonii f. *dawsonii*, 164

Mexipedium xerophyticum, 233
Paphiopedilum armeniacum 233
hand pollination, 87, 119, 120, 175, 176, 194, 213, **227**, 233, 240, 257, **272**, 290
harvesting, 35, 151, 200, 225, 234, 273–4, 277
immature, 120, 237, 274
number, 34, 35
  in a single capsule of *Orchis maculata*, 34
  in a single capsule of *Grammatophyllum speciosum*, 35
packets, 38
pods (see capsules)
propagation, 86, 182, 198, 222, 234, 237, 241, 257, 265, 274, **275**, 283
recalcitrant, 84
sowing, 38, 58, 181, 270, 274, 275, 281
storage, 120, 140, 200, 225, 234, 268, 270, 274, 277 (see also, banks)
viability testing, 120, 143, 270, 275, 276, 277
Seed for Life, 277
seedling(s), 19, 86, 117, 143, 175, 176, 180, 181, 208, 223, 237, 240, 263
  conditions for survival, 22, 175, 223, 243, 252
  establishment in greenhouse cultivation, 23, 181, 198, 225, 237
  establishment in sterile culture, 23, 180
  fungal symbionts, 79, 91, 144, 180, 181, 223, 238
  genetic diversity of, 265
  growing for profit, 231
  growth *in vitro*, 87, 147, 151, 175, 181, **183**, **184**, 198, 199, 263, 272, 273, 274, 283, 291
  identification using DNA, 100
  natural establishment, 199, 223, 225, 239, 253
  of *Angraecum longicalcar*, 79
  of *Aerangis ellisii*, 82
  of *Bletia urbana*, 143
  of *Cattleya labiata*, 108
  of *Dichromanthus aurantiacus*, 143
  purchase of, 23
  recruitment, 223, 231
  reintroduction of, 24, 208, 222, 223, 225, 233, 238, 243, 252 (see also, reintroduction)
  spontaneous, 20, 21, 35, 156, 238
  survival rate, 143, 144, 177
Seehawer, 207
senile population, 21
sepals, 32, 105, 126
Serapias,
  S. lingua, 154
  S. parviflora, 154
sexual parasitism, see pseudocopulation
sexual reproduction, 28, 98
shade house, 84, 120, 139, 199, **201**, 213, 222, 256
sheep, 185
Shephard, Andy, 215
Sheehan, Marion, 271
Shifting Baseline Syndrome (generational amnesia), 188
ships, 55, 66, 135, 152, 264, 280
  HMS Erebus, 152
  HMS Terror, 152

Singapore, 36, 113, 222, 250, 251, 252, 253, 276
  Bukit Batok Nature Park, 252
  Bukit Timah Nature Reserve, 251
  Pasir Ris Park, 253
  Singapore Botanic Gardens, 95, 190, **191**,192, 253
  Tengah Forest, 95
Site of Special Scientific Interest (SSSI), 244
Skinner, George Ure, 48, 49, 144
Skutch, Alexander, 28, 138
slash and burn farming, 75
slide mounts, 38, **78**
slugs (snails), 238, 273
Small, John, 223
Smith, Zöe, 50, 90, 91
Smithsonian Environmental Research Center (SERC), 75, 243, 244
Smithsonian Gardens, 257
Smithsonian Institution, 243, 244
snakes,
  brown snake (*Pseudonaja textilis*), 50
  cottonmouth (*Agkistrodon piscivorus*), 19
  fer-de-lance (*Bothrops asper*), 153
  venomous, 19, 49, 71, 153
snow, 158, 243, 280
soap, germicidal, 58, 100
Sobralia, **289**
social media, 146, 292
sodium chloride (salt), 283
soil, 37, 38, 91, 106, 134, 143, 144, 154, 155, 156, 172, 177, 180, 182, 193, 210, 223, 238, 243, 245, 246, 257, 280, 284
  loamy, 243
  nutrient-poor, 176
solar energy, 82
Sophronitis coccinea (syn. *C. coccinea*), **170**, 171
Soto, Miguel, 103, 163
South Africa, 32, 120, 190
  Cape Floristic Province, 190
  Cape Institute of Micropropagation, 120
  Darling, 190
South America, 8, 34, 42, 49, 63, 66, 102, 134, 135, 155, 158, 171, 189, 192, 198, 200, 217, 231, 257, 288
South Carolina, 271
  Clemson University, 271
South China Sea, 129
Southern Hemisphere, 210, 222
Spain, 138, 154
Spanish conquistadore(s), 159, 280
Spanish moss (*Tillandsia usneoides*), 260, 291, **292–293**
Spathoglottis, 35
Specklinia colombiana (syn. *Acostaea colombiana*), 208
Spence, Phil, 72
sphinx moth (*Tinostoma smaragditis*), 69
spiders, 14, 253
spores, 23, 35, 156, 275
Sprunger, Samuel, 235, 246, 249
Sri Lanka, 278
Stanhopea, 42
  S. lietzei, **173**, 175
Stelis, 215
Stice, Andy, 77, 79
stigma, 30, 200, 230, 240
stomata, 39
Stoneman Douglas, Marjorie, 227
Straits of Magellan (Elizabeth Island), 102

students, 9, 11, 19, 22, 23, **31**, 35, 68, 86, 100, 101, 102, 142, 143, 175, 192, 193, 194, 196, 197, 200, **201**, 215, 223, 225, 230, 235, 244, **265**, 271, 278, 283, 291, 292
Stuppy, Wolfgang, 268
Suarez, Iris, **142**
subterranean, 83, 155
sugar(s), 28, 30, 31, 37, 38, 39, 42, 116, 182, 283
  cane, 171, 259
  fructose, 29
  glucose, 29
  sucrose, 29
Sugii, Nellie, 69
Sulyok, József, 239
Sumatra, 35, 70, 234
sun birds, 32
SUV(s), 79, 80, **81**
Swainson, William, 105, 106, 107
Swarts, Nigel, 277
Sweden, 283
Swift, Jonathan, 159
Switzerland, 145, 235, 240

T
Taiwan (Formosa), 11, 63, 113, 124, 129, 131, **161–162**, 167, **168**, 170
taxonomy, 9, 98, 99, 103, 162, 181
tea, 119, 222
  teabags, 274
Teagueia, 215
Telipogon, **157**, 158, 213, 280, 281, **282**
  photography, **281**
  T. diabolicus, 95
temperature(s), 57, 62, 63, 66, 131, 143, 152, 153, 155, 156, 162, 164, 180, 196, 199, 207, 234, 243, 250, 270, 272, 280, 284, 285, 292
Tepoztlán Orchid Society, 164
termites, 83, 172
terrestrial, 32, 34, 35, 38, 39, 42, 67, 72, 82, 83, 85, 98, 119, 143, 145, 153, 156, 158, 159, 176, 180, 182, 185, 190, 192, 200, 207, 210, 214, 223, 230, 250, 271, **272**, 276, 277, 283, 284
testa (seed coat), 34
testicle, 116
tetraploids, 101, 245
tetrazolium (2,3,5 triphenyl tetrazolium chloride, TZ), 198, **268**, 276
Thailand, 36, 74, 112, 113, 276, 278, 279
  Bangkok, 278
  Mahidol University, 278
  Prince of Songkla University, 35
Thammasiri, Kanchit, 278
The Orchid Thief, 17
Thompson, Oscar, 14
Thompson, Peter, 234
Thoreau, Henry David, 153
thrips, 32, 264
Tibet, 158
Ticos, 112
tiger(s), 35, 74, 112, 251
timber (see logging)
tissue culture, 23, 117, 151, 200, 225
toads, 218
tobacco, 53
Tobar, Alvaro, 196
Tobar, Francisco, **281**
Totonacs, 158, 159
toucans, 63, 217, 218

tourism (see ecotourism)
TRAFFIC, 119
trampling, 126, 185, 210
translocation (transplantation), 86, 87, 194, 222
trees,
   Araucaria angustifolia, 171
   black cedar (*Juglans neotropica*), 207
   caracolí (*Anacardium excelsum*), 260
   Cecropia, 207
   Cedrela fissilis, 172
   Erythrina poeppigiana, 189
   guácimo colorado (*Luehea seemannii*), 260
   guano prieto (*Acoelorraphe wrightii*), 106
   in Taiwan,
      Taiwania cryptomerioides, 167
      Taiwan cypress (*Chamaecyparis formosensis* and *C. obtusa* var. *formosana*), 167
      Taiwan red pine (*Pinus taiwanensis*), 167
      Taiwan spruce (*Picea morrisonicola*), 167
      Rhododendron formosanum, 167
      Pinus armandii var. *mastersiana*, 167
   pop ash (*Fraxinus caroliniana*), 24
   ironwood (*Chionanthus caymanensis*), 139
   Magnolia polyhypsophylla, 208
   oak (*Quercus humboldtii*), 207
   Polylepis, 280
   pond apple (*Annona glabra*), 24
   rain tree (*Samanea saman*), 252
   Robinia pseudoacacia, 235
   tachuelo berrugoso (*Zanthoxylum verrucosum*), 260
   wax palm (*Ceroxylon quindiuense*), 207
Trevora, 218
Trudgill, Dave, 245
truffles, 85
trunks (see bark)
Tschiffely, Aimé, 288
tubers, 38, 82, 91, 116, 119, 120, 156, 284
Tucker, John, 162, 163
Tulasnella, 143
   T. inquilina, 271
type specimen, 84, 102, 103, 210

U

UAMH Centre for Global Microfungal Biodiversity, 285
UK Darwin Initiative Project, 119, 276
UK National Orchid Collections, 277
underground orchid, 83
United Kingdom (UK), 59, 98, 99, 113, 119, 126, 140, 145, 150, 151, 152, 153, 154, 162, 163, 164, 171, 177, 180, 196, 197, 200, 244, 245, 246, 260, 261, 263, 270, 277, 279, 290, 292
   Birmingham, 58, 162
   Chelsea, 50, 57, 291, 292
   Chiltern National Landscape, 140
   Dorset, 154
   Down House, 9
   Eades Meadow, 244, 245
   England, 9, 55, 57, 59, 112, 119, 126, 153, 154, 162, 163, 234, 235, 245, 264
   Essex, 154
   Exeter, 57
   Gait Barrows National Nature Reserve, 238
   Glasgow, 105, 107
   Gloucestershire, 245
   Herefordshire, 245
   High Wycombe, 140
   Jersey, 260
   Kent (Samphire Hoe), 36
   Kidderminster, 196
   Lancashire, 238
   Lincolnshire Wolds, 152
   Liverpool, 58
   London, 48, 50, 52, 57, 58, 119, 126, 140, 151, 154, 264, 268, 291
   Scotland, 245
   West Midlands, 245
   Worcestershire, 244, 245, 246
   Yorkshire, 235
UN Global Biodiversity Framework, 206
UN Summit on Biodiversity, 289
United States of America (USA), 149, 151, 153, 155, 162, 164, 188, 197, 227, 257, 260, 272, 288, 291
   Maryland, 243
universities,
   Mahidol University, 278
   Prince of Songkla University, 35
   Universidad Nacional Autónoma de México (UNAM), 143
   Universidad Veracruzana, 166
   University of Cuenca, 95, 279
   University of Glasgow, UK 105
   University of Kitwe, Zambia, 120
   University of Lancaster (The Fylde College), 200
   University of Leeds, 237
   University of Mexico, 50
   University of Western São Paulo (UNOESTE), 199, 200
   Utrecht University, 244
urban heat island, 140, 249
Uruguay, 171
US Fish & Wildlife Service, 19

V

Vacherot and Lecoufle, 124
Vanda, **40–41**, 73, 113, 114
   V. falcata (syn. *Neofinetia falcata*), 113, **115**
   V. sanderiana, 71–72, **73**
   V. tricolor var. *suavis*, 198
vandalism, 82
Vanilla 36, 106, 117, 119, 158, 159, 200
   V. barbellata, 155
   V. planifolia, 117, **118**, 119, 158
Vasquez, Yanny, 275
Vavilov, Nicolai, 268
Veitch, James, 48, 57, 105, 214, 234
Venezuela, 63, 189, 206, 290
   Aricagua, 189
   Caracas, 189
   Cerro El Ávila, 189
   Mérida, 189
   Portuguesa, 189
   Tapo-Caparo National Park, 189
velamen, 39
Victoria, Australia, 86, 87, 90
Vietnam, 95, 124, 230, 276
virus, 222, 263, 264, 265
   Cymbidium mosaic virus (CyMV), 259, 264
   Odontoglossum ringspot virus (ORSV), 264
volcano(es), 140
   Antisana, 280
   Chimborazo, 280
   Krakatoa, 35
   Pichincha, 94
   Soufriere Hills, 140
   Tungurahua, 215
   Xitle, 143
volunteers (also see citizen scientists), 86, 190, 197, 225, 238, 239
   Victoria, Australia, 87, **90**
voucher specimen, 100, 120, 273

W

Wakehurst, Sussex, 35, 234, 237
wallabies, 83
Wallace, Alfred Russel, 70, 74, 167, 251
Wallace's Line, 74
Wallis, Gustave, 48, 50, 105
walnut, 58
Warcup, Jack, 85
Warner, R., 58
Warren, Richard, 135, 171, 175, 176, 177, 207
wasps, 34, 83, 159, 172
weedy orchids, 36, 37
Weir, John, 51
Wells, Terry, 185
wetlands, 20, 190, 209
Whigham, Dennis, 38, 243–244
Wightman, Nicholas, 120
wildlife trust(s), 234, 244, 290
Willems, Jo, 244
Williams, B. S., 56
Wilson, Ernest H., 48, 131, 279, 280
Wilson, E. O., 152, 162
Withner, Carl, 256
Wimber, Don, 101
woodland, 145 185, 192, 208, 235, 239, 243, 246
World Orchid Conference, 272
World Wildlife Fund (for Nature), 85
Wright VI, John Parke, 291
Writhlington School Orchid Project, 196

Y

Yachang Nature Reserve, 116, 130, 131
Yam, Tim Wing, 250
yellow fever, 49
Yokoya, Kazutomo, 79
Yumbo Trail, 213

Z

Zambia, 119
   University of Kitwe, 120
Zandoná, Luciano, 11, 172, **173**, 200, 222, 223, **286–287**
Zettler, F. William, 263
Zettler, Lawrence, 75, 77, 79, 234, 271
Zoological Society of London, 139
Zygopetalum maxillare, **27**